THE GOLDEN INSECT

THE GOLDEN INSECT
A handbook on beekeeping for beginners

Stephen Adjare

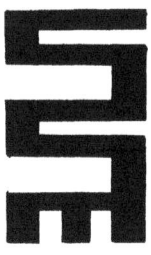

Technology Consultancy Centre, University of Science and Technology, Kumasi *in association with* IT Publications 1984

Published by ITDG Publishing
103–105 Southampton Row, London WC1B 4HL, UK
www.itdgpublishing.org.uk

© TCC and Intermediate Technology Publications 1981, 1984

First published by TCC in 1981
Revised edition published in 1984
Reprinted 1989
Print on demand since 2003

ISBN 0 946688 60 5

All rights reserved. No part of this publication may be reprinted or reproduced or utilized in any form or by any electronic, mechanical, or other means, now known or hereafter invented, including photocopying and recording, or in any information storage or retrieval system, without the written permission of the publishers.

A catalogue record for this book is available
from the British Library

ITDG Publishing is the publishing arm of the Intermediate Technology Development Group. Our mission is to build the skills and capacity of people in developing countries through the dissemination of information in all forms, enabling them to improve the quality of their lives and that of future generations.

Printed in Great Britain by Lightning Source, Milton Keynes

Contents

	Page
FOREWORD	ix
ACKNOWLEDGEMENTS	xi
INTRODUCTION	1
CHAPTER ONE: EQUIPMENT	12
The Beehive	12
Tanzanian Top Bar Hive	14
The Long Transitional Hive	14
The Smoker	16
The Veil	17
The Beesuit, Gloves and Shoes	18
A Hive Tool and a Knife	18
A Brush or Quill	18
A Swarm Catcher	18
A Match Box	18
CHAPTER TWO: THE COLONY	19
The Drone	19
The Queen	20
The Worker	21
Cleaning Operations	23
Ministering to the Queen	23
Orientation Flight	24
Ventilating the Hive	24
Guard Duty	25
Executions	25
Execution of Drones	25
Robber Bees	26

The Field Bees	26
Impregnating the Queen	27
The Unfertilized Queen	29
The Queenless Colony	29

CHAPTER THREE: HIVING A COLONY — 31
The Swarming Season	31
The Entrance of the Hive	34
How to Capture a Swarm	35
Hiving the Bees	36
Preparing the Hive	38
Forcing Bees Out	40
Hanging the Hive	41
How to Feed Bees	41

CHAPTER FOUR: SOME COMMON PRACTICES IN APIARY MANAGEMENT — 44
Controlling Swarming to Your Advantage	44
Hiving by Dividing an Established Colony	47
How to Unite Bees or Make an Increase	49
Uniting a Swarm and a Colony	50
Uniting a Queenless to a Queen-right Colony	50
Formation of a Nucleus	51
When a Colony Swarms	51
How to Prevent Robbing	51
Feeding Bees	52
Leave Some Honey for the Bees	53
Unconscious Method of Feeding Bees	54
Watering Bees	54
Keeping Records	56
Colony Records	56
Operational Records	56
Other Managerial Practices	57

CHAPTER FIVE: HOW TO MANIPULATE BEES AND EXTRACT HONEY AND BEESWAX — 58
Why do you have to Handle Bees?	58
Honey Harvesting and What to Look For	58
Other Signs to Note	60
How to Harvest Honey and Control Brood Nest	61

Bee sting 64
On the Attack 65
Avoid Stings 66
Extraction of Honey 66
   The Solar Wax Melter 67
   Traditional Method of Extracting Honey and Beeswax 67
   Beeswax Extraction 68
   Moulding the Beeswax 71

CHAPTER SIX: FACTORS MILITATING AGAINST THE BEE INDUSTRY 73
Natural Climatic Conditions 73
Natural Enemies and Pests 74
   Ants 74
   The Wax-moth 75
   Acherontia Atropot 76
   The Lizard 77
   The Bee Pirate 78
   The Hive Beetle 78
   The Alpine Swift Bird 79
   Other Organisms 79
   The Bee Louse 79
   The Bee Scorpion 79
Bee Friends 79
Human Activities 80
   The Honey Hunter 80
   Bush Burning 80
   Bee Burning 81
   The Palm Wine Tapper 81
   Poisoning Bees 82
Other Miscellaneous Problems 83
   Queenlessness or Unfertilised Queen 83
   Absconding Colony 83
   The Spider's Web 84

CHAPTER SEVEN: SOME DISEASES OF THE HONEY BEE 85
The Brood Diseases 85
   The American Foul Brood 87

The European Foul Brood	88
Stone Brood	89
Nosema-like Protozoan	89
The Chilled Brood	90
Bald Brood	90
Genetic Faults	91
Diseases of the Adult Bee	91
Nosema	91
Dysentery	92
Paralysis	92

CONCLUSIONS 93

APPENDICES
A: The Uses of Honey	94
B: The Uses of Beeswax	96
C: Physical Properties of Beeswax	101
D: Honey Bee Forage Trees in Ghana Forests	102
E: Kenyan Top Bar Hive, Ghana Version	104

Foreword

This manual on beekeeping in tropical conditions has developed from the author's work at the Technology Consultancy Centre (TCC) of the University at Kumasi, over the last ten years. When the TCC first became interested in beekeeping in 1977, most of the little honey and beeswax sold in the markets of Ghana came from wild honey tappers. These men used fire to drive away the bees from their nests and this resulted in destruction of the brood combs and poor quality honey of smokey flavour. So a decision was taken to try to benefit from the experience of Kenya and Tanzania where beekeeping is a widespread rural industry and the products of the honey bee are available in abundance for home consumption and for export. In this endeavour the TCC was greatly helped by the Commonwealth Secretariat of London. The TCC also proceeded hand-in-hand with the Forest Products Research Institute (FPRI) of CSIR in a programme aimed at identifying suitable beekeeping areas in Ghana.

Soon bee hives were being produced to the Kenyan pattern by the University's Department of Building Technology, by FPRI and by several private carpenters. The TCC began to produce protective clothing and sub-contracted the manufacture of smokers to a local small industry. Pilot apiaries were set up at the University's Botanical Gardens, at Atebubu in Brong Ahafo Region, at Accra and at Hohoe in the Volta Region. This activity culminated in a National Beekeeping workshop held by the TCC in Kumasi in January 1981. It was then realised that a great deal of interest had been generated and people wanted to start beekeeping in almost all parts of the country. A Beekeeping Association was founded and a quarterly newsletter

(Ghana Bee News) introduced. The TCC began to plan two-week practical training courses for beekeepers and the need for a handbook or training manual became apparent. This book was produced to meet this need.

Neither Mr Adjare nor the TCC claim to have any expert knowledge of apiculture. However, Mr Adjare has had several years practical experience and he has read very widely on the subject. We have also benefited greatly from the advice of visiting experts, the most notable of whom is Mr Curtis Gentry of the USA. Much valuable knowledge has been gained from the International Bee Research Association of the UK and from visits to Tanzania and Kenya. On this basis this book is founded. However, it is written by a Ghanaian for Ghanaians — by an African for Africans. It takes account of the socio-cultural heritage of the people and is written in language they can understand and uses symbols they can recognise. If this leads the author away from standard English at times it is hoped that the purists will forgive us. The aim is to put into the hands of the Ghaniaian and African beekeeper information that he can readily understand and put to immediate use. We hope that we have been reasonably successful at this first attempt. No doubt future editions will make good any errors or omissions.

The last word must go to our primary reader, the Ghanaian beekeeper. We wish you well in all your endeavours to promote your cottage industry. It is hoped that you will turn to the TCC for any further help or advice that you may need. Good luck.

John Powell
Director TCC
Kumasi
Ghana

Acknowledgements

This world of ours is a place where no individual can boast of providing everything for himself without assistance from others. In publishing this book, the writer could not have done it all alone. *The Golden Insect* therefore is the brain-child not only of the writer and his editor but of a number of people whose kindness and benevolence must be acknowledged.

First and foremost my heartfelt thanks go to the General Manager and staff of the Intermediate Technology Industrial Services of Rugby, for funding the printing of 1,000 copies of the first edition of *The Golden Insect* and sending them to us in Ghana. They have again funded the printing of this second edition. Mr Howard Veatch of Dadant and Sons, Illinois, must be thanked for the use of his pictures and permission to use portions of their book *The Hive and the Honey Bee*. The lists of the uses of honey and beeswax which appears in the Appendix are theirs.

My next thanks go to Dr. Eva Crane of the International Bee Research Association for her contribution in allowing me to use the mating queen-bee photograph: a picture which has not been found in any other beekeeping manual. Mr B. Clauss of Botswana also must be thanked for his pictures and text books sent to the Centre free of charge.

Let us now come home to Ghana and thank one group which has been the backbone of our bee promotion. It was the first to send a contingent of 25 volunteers to attend the first National Workshop held in Kumasi between 28 and 31 January 1981. In 1983, the same group sent a similar number of participants; it also sent two renowned apiarists in the persons of Curtis Gentry and Michael Culp from far away in the United States to add their authority to the lectures of the bee schools. In fact the organization carries apiculture in Ghana on its back as a mother does her baby; it is the US Peace Corps in Ghana and the man behind the scene is Mr Ross Kreamer, the Associate Director for Agriculture. The writer and the entire staff of the TCC register their appreciation of the Peace Corps' enormous contribution.

I have not forgotten the good work that Mr and Mrs Ralph Moshage are doing at the TCC. Mrs Marlene Moshage needs special praise for assisting in the making of this second edition of *The Golden Insect*. All the nine photographs found in Chapter 5 concerning the extraction of beeswax are hers. Indeed Marlene has unceasingly assisted in the promotion of beekeeping in Ghana and her name will be printed boldly in the bee history of this country. In all, *The Golden Insect* salutes the US Peace Corps for giving its fullest support to beekeeping; not forgetting its effort in mounting a series of village campaigns to educate the Ghanaian farmer on the need to process beeswax.

Our sincere thanks also go to Mr W.A. Osekre and Mr Bonful, University photographers, who printed our photographs for this manual.

Mrs Comfort Adjare, Master A.K. Opoku and Akosua Tiwaa must be congratulated for the part they played in the extraction of beeswax. Mr Addo-Ashong, Director and Mr F.A. Awuku of the Forest Product Research Institute should not be forgotten for providing the common and local names of the bee forage trees.

Lastly, thanks go to Miss Esther Akom, the secretary of the Director, TCC and her assistants Miss Emma Achampong, and Miss Grace Dery. These three ladies typed the draft of this second edition. To these and to all who in diverse ways assisted in the publication of this edition, Stephen Opoku Adjare says thank you.

Bee Population of West Africa

Introduction

Honey has, through history, been one of the chief sweet foods of man. The history of the Israelites (Genesis 43:11) confirms that Jacob asked his sons to carry, among other gifts, some little honey to be given to 'the man' (Joseph). Honey has been and is still collected from hollow tree trunks, abandoned ant hills and from crevices. It is delicious and compared to sugar, it is rich in nutrients.

This valuable food is still collected in a crude manner all over West Africa during the burning season at the time when farmers are preparing the land for farming. The method of collection (except for a few places in Guinea Bissau and Senegal) is, even at the time of writing, barbaric and wholly unacceptable. Honey hunters, who discover a nest of honey bees, either hanging on a branch of a tree or in a tree trunk, prepare a torch with dried grasses or palm branches and smoke out the bees. In the process, hundreds of these precious insects are burnt to death. This practice, in East Africa, is likened to a shepherd who has to kill his animal in order to milk it. In Senegal and Guinea Bissau, where beekeeping has long been a traditional practice, bees are not killed in this way as they are in other African countries, including Ghana. The traditional honey hunter in Ghana can safely be called a 'bee killer'. After burning some of the bees and rendering them less aggressive, he uses chisels, axes, machetes and mallets to force open the nest. He then scoops out all of the contents hurriedly without leaving the brood nest behind and carries his booty home. A colony of bees treated in this way is overwhelmed and the only option left to the few that survive this cruel ordeal is to abscond and find a new nest elsewhere. The old nest is completely destroyed, thus reducing the number of hollows where bees can conveniently make their nests. The practice is still going on without the slightest governmental restriction since honey hunting is the major source of honey supply in this sub-region. The barbaric harvesting method also accounts for the impossibility of harvesting a colony more than once annually since the bees have to re-establish their nest in a new location.

Now it appears that hollows in tree trunks have decreased so much that many colonies, not finding such natural homes, have resorted to making their nests in branches of trees. In the transitional forest zones and the savanna woodlands, where bees are abundant, a swarm of bees that decided to make their permanent home in tree branches would probably choose the silk cotton (kapok) or the giant *baobab*. These trees contain large quantities of water and so they are not consumed even if all other trees are burnt during an outbreak of bush fire in the dry season.

In fact, this crude method of honey harvesting itself accounts for a number of the fire outbreaks in the grasslands and the forests as honey hunters often throw their burning torch carelessly into the dry bush after harvesting.

Why do people burn the bush during the dry season? I am not qualified to enumerate all the reasons why savanna grasses are burnt by the inhabitants who know perfectly the dangers of wild fire outbreaks. For the sake of beekeeping the most important obstacle that concerns us is bush-burning in the very densely bee-populated areas. According to a government officer at Amanteng in Brong Ahafo Region, bees are deliberately burnt in order to reduce their numbers to enable the inhabitants to live peacefully. This statement, whether justified or not, is illustrated by the Accra *Mirror's* front page article on Friday, 19 December, 1980 headlined 'Bees on the Attack'. This episode happened at Dormaa-Ahenkro in the Brong Ahafo Region during the opening ceremony of the Kwafie festival *durbar,* and the statement continued: 'Chiefs clad in their beautiful festival finery had to run for dear life abandoning their palanquins . . .'

Another catastrophe which appeared on the front page of the *Daily Graphic* of the 24 December of the same year (barely one week after the Dormaa-Ahenkro tragedy) was headlined: 'Mourners attacked by bees at Vakpo'. Those attacked included choristers from Tema and Accra and one goat was stung to death. The article continued that 'concerned people in the town are seriously discussing effective measures to wipe out the bees which had been making unprovoked raids on passers-by'. There are many other bee attack cases where human lives have been lost. Atebubu District lost two children in 1979 and two in December 1980. Damango has also suffered losses. Similar unpleasant attacks contributed, in this writer's view, to the decision by agricultural authorities of the First Republic of Ghana to reject the indigenous tropical honey bees, *apis mellifera adansonni,* in favour of caucasian exotic bees from Europe which are less aggressive. Alas, those bees, imported with scarce foreign currency, perished almost immediately and the development of apiculture in Ghana was abandoned indefinitely.

Bees are present all over the country but the savanna woodland and the transition forest have more bees than the forest and the desert oasis areas. Bees at Kumasi (in the deciduous forest) do not seem to be so aggressive as those in the grassland while those found near the desert are more temperamental than those in the grassland — the hotter the climate the more aggressive the bees seem to be.

So far as the two bee invasion episodes referred to above are concerned, the bees might have been disturbed by the noise of singing, drumming, dancing and merry-making or might have been attracted or incited by a special sweet-scented perfume or cosmetic which probably contained beeswax. They tend to attack furiously anybody in dark clothes.

They react violently to talking and noise-making in general. Therefore, it is again advised that anybody approaching a hive should advance quietly and stealthily to avoid the fury of the bees.

Please, my good bee enthusiast, do not be alarmed by the danger of the bees. By nature they are not aggressive all the time. All you have to do is to avoid them during the hot time of the year or the hot time of the day. If, by chance you are stung by any bee, please cover the spot and run away quickly. Your enemy has left some pheromone, which is a powerfully scented chemical, on your skin which summons more angry bees to join in the attack. After removal of the sting, apply water or some liquid from a strong-smelling herb to drive away the scent. This will allow you to go on your way without further attack.

Our tropical bees are intelligent and highly industrious although they produce less honey than the *apix mellifera* in the temperate zone. However, the African bee is a superior producer of wax. Whilst the *a.m. adansonii* works all the year round, because of the tropical climate, European bees only work during the summer months and over-wintered (kept indoors) throughout the cold season. To take advantage of the favourable weather and the natural industriousness of the bees, we must find good homes (i.e. beehives) for them and provide water as well.

Between the months of December and May, there is little or no water in the streams and rivers and the bees have to resort to other means to get water. They may claim ownership of any water collected by man for household and drinking purposes. They will also collect water from any moistened area on the ground or from any wet cloth or objects. On several occasions, I have watched them collect water in this way, between January and March, especially during the peak of the dry *harmattan* period. Therefore it is advised that any beekeeping project should make provision for regular water supply for the people as well as for the bees. This will help to eliminate most of the brutal assaults by bees which make people hesitate to approach a swarm or any bee colony.

Bees work very well during the warm periods of the day. They drink large quantities of water and need water to cool the hive during the dry season. The *harmattan* season between December and February is the time bees cart large quantities of water from obscure places like the urinal, wet portions of river banks, swamps and lakes. Nowadays, the swampy areas in our forests are giving way to dry soils. Desertification in Ghana is a serious problem facing those interested in the apicultural development of our sub-region.

Those Ghanaians who live in the regions of southern Ghana have noticed that for the past three to four years, the annual rainfall has decreased alarmingly. The swampy forest paths are no more. Large rivers in Ashanti Akim, for example, which have never failed to flow throughout the year are now nothing but dust basins. Farmers, lumber-men and all land utilizers must be blamed. Farmers have cleared all the bushes along the banks of our rivers which exposes the waters to direct sunlight and increases the rate of evaporation and soil erosion. All the catchment areas which were not utilized by our great-grandfathers have been cleared which causes the water tables to drop considerably. Now the famine and drought which have brought untold suffering to the Ghanaian population at the time of writing is the result of careless utilization of the forest. Forests which nature took millions of years to nurse, man has destroyed within decades. The earlier the government takes action to implement a giant re-afforestation programme to clothe the banks of all our rivers, streams and catchment areas with high nectariferous trees or fruit crops like mangoes, citrus orchards, *etc.* the better. It is hoped that this problem may be taken over by the forestry department which can work side-by-side with beekeepers so that honey yields in this region will greatly improve to the advantage of all. By instituting this corrective plan, many large apiaries could be sited in these man-made forests. The bees will be fed and watered whilst man may take advantage of the fruits as well as the honey that will be produced.

Why should we keep bees?

Many Ghanaians have lost hope in obtaining help from their hospitals because the dispensary shelves are empty. Many factories and manufacturing industries are idle because they have had no import licences to purchase inputs of raw material from abroad. Many craftsmen and factories have stopped production because they cannot get beeswax. And above all, many farmers are harvesting less from their fruit crops because there are not enough bees on their farms to provide adequate pollination. These are but a few of the reasons for promoting beekeeping. Above all, honey is sweet; honey is money. The value of the honey bee can be seen by reading Appendices A and B of this booklet. The uses of honey in drug-making, of beeswax, which is a multi-purpose industrial raw material, of propolis, pollen, bee milk and bee venom all convince us that the honey bee is really one of the most useful creatures of those which man has domesticated. West Africa is lucky to have this useful and industrious insect in large quantities and its potential for commercial production of multi-flora honey is certainly under-utilized. I have tasted different types of honey from many countries and I rate ours as among the sweetest. However, I keep my fingers crossed when commenting on the general quality because the methods being employed here by our traditional suppliers (namely, the honey hunters) render the commodity unacceptable by international standards.

We must express our thanks to Ralph Moshage and Mike Culp for assisting the TCC in introducing the solar-wax extractor which beekeepers in the savanna can effectively use to extract both honey and beeswax. It is hoped that this will go a long way to make clean honey and wax available for home consumption and for export.

Beekeeping is being stimulated in Ghana at a time when its economy is on its knees. Inflation is at a peak. The labourer who receives a 40 cedis[1] daily wage buys a 0.25kg loaf of bread at 60 cedis and 1kg of beef at a controlled price of 83 cedis! Production in every sector of the economy has fallen except, of course, in the manufacturing of coffins and other necessities for the dead. There is real hunger in the land and belt-tightening is the order of the day. Many factories have closed down and the few working are operating at under 10 per cent capacity. There is frustration all over the country. Citizens of the land who migrated to Nigeria and other sister countries are being driven back to face realities in their homeland. Why should such a catastrophe occur in a country that abounds in rich natural and human resources? Again, I am not well equipped to offer detailed answers but one reason concerning us is that most enterprises use complicated, sophisticated, imported machines and raw materials. Hence, broken-down machines lack spare parts which the country's scanty foreign reserves cannot afford to replace during this period of depression. Therefore proper investigation and research must be made to utilize available resources and talents. As there is rich apicultural potential in our sub-region it is our responsibility to make use of this natural abundance.

In our tropical Africa, beekeepers do not necessarily have to

---

1. Current rate of conversion to sterling: approx £1 = ₡50 (50 cedis).

purchase honey bees to start beekeeping as is done elsewhere. Even in the forest where we do not recommend a giant commercial beekeeping venture, a beehive properly baited takes an average of four weeks to be colonised during the swarming period. The individual beekeeper does not need to import any foreign equipment. He can obtain all necessary bee-equipment from within the country.

In summary, the economic value of honey, beeswax, propolis and the honey bee's pollinating activities which improves production of our cash crops like citrus fruits, coffee, cola, sheabutter and food crops cannot escape us. In our bee industry we should not make any mistake by placing honey production first. We should give priority to pollination especially in our cocoa industry as well as on our citrus farms by populating our farms with bees. I would therefore, humbly appeal to all our farmers to consider this.

The need to promote beekeeping was realised by the Technology Consultancy Centre in 1975. This was the result of coming into contact with traditional brass founders living on the outskirts of Kumasi at Kurofofurom.

These craftsmen use beeswax as one of the principal materials for casting (in the lost wax process) their statuettes which they make for the tourist trade. At that time beeswax was very scarce, hence the price was between 60 and 80 cedis[1] per kg. Thus Mr Kwesi Opoku-Debrah, the then officer in charge of rural crafts at the TCC who made the preliminary survey to determine potential of the bee industry, writes, referring to the scarcity and high prices of beeswax: "this prompted the TCC to find a permanent solution to their (the brassfounders') dilemma as part of exploiting and upgrading their industry".[2]

Presently a few tribes in isolated areas keep bees in West Africa. In the Northern and Upper Region of Ghana and Upper Volta pots are used as beehives; but large-scale honey production by an individual or any institution is unknown. However, these locations are among the best areas in the region where we are advising interested parties to go into beekeeping and beeswax collection on a commercial scale.

1. The official rate of the US dollar was 1.15 cedis.
2. 'A report of an exploratory study of Beekeeping in Tanzania and Kenya and its potential in Ghana' by K. Opoku-Debrah and E. Osei Bossman.

Most of the honey consumed in the region is brought to the market by honey hunters whose method of honey collection has been mentioned above. After squeezing the honey from the combs most honey-hunters throw away what remains which contains the precious wax (which Ghana has started collecting through the TCC and Mansusu Agencies). Nigeria also collects beeswax but it is not known whether it passes through the government's official channel. Reports collected by the writer from village honey hunters and an *Alhaji* at Kano indicates that the 'dirty' stuff is packed in barrels and flows through Kano Airport. Thus not only the brassfounders but other craftsmen and several industries including Seward UAC Ghana, Lever Brothers, Johnson Wax and many cosmetic, pharmaceutical and textiles industries lose this precious local material. The nations also lose a potential foreign exchange earner while consumers have to spend the little foreign currency they get to purchase many tons of beeswax to feed the local manufacturing industries.

Beeswax is a multipurpose industrial material (see Appendix B) needed all over the world but which requires very little investment. In fact most of the work required has been completed by the honey bee. Considering the quantity of honey that the bee converts into wax is in the ratio 15:1, beeswax should be regarded as a precious and very expensive commodity. In November 1982, a bottle of honey (about 1kg) was sold in Accra for 350 cedis. In effect, if 1kg of honey cost 200 cedis (an average price) then a kilogram of beeswax should consequently cost 3,000 cedis.

However, in practice, the price of beeswax does not reflect this as the selling price is determined by demand, which in its turn depends upon the uses which society makes of it.

The annual yield of honey depends on the management of the hives. In Accra, a colony of bees that entered a beehive on the 5 April, 1983 produced 11kg of honey in two months (i.e. on the 5 June 1983). Beekeepers in Ghana have reported their first harvests as follows:

                Northern Region 23-32kg
                Brong Ahafo   4- 6 gallons (28-42kg)
                Ashanti   20-30kg
                Greater Accra   25-30kg

In Kumasi, one wild honey nest when harvested yielded two kerosene tins of honey (i.e. 60kg). Thus these figures show that beekeeping can be a profitable venture considering that:
   i. the industry is new and that better harvests are to be expected in the second year.
   ii. the year (1983) was a bad one for the honey bees. There were several outbreaks of wild bush fires, wind storms and drought which destroyed many colonies of bees.
   iii. all the beekeepers were new to the industry and inexperienced as managers. Several colonies, for example, were either harvested prematurely or late and therefore did not produce maximum yields.
   iv. waste of honey could be avoided by the use of the hand to squeeze the honey comb to get the honey out. With the introduction of the solar wax extractor, it is hoped that honey extraction will be more efficiently and easily accomplished.

It is unfortunate that up to the time of writing, no statistics have been received for extraction of beeswax but we assure beekeepers that this should be about 6-7 per cent of the honey yield. We further advise that anybody who wants to go into commercial beekeeping should begin it in the following vegetational zones:
a. the transition forest
b. the savanna woodland
c. the Sudan savanna
d. the coastal scrubland

The key plants to look for (see Appendix C) must include: mangoes *(mangifer indica)*, acacia, parkia *(dawadawa)*, eucalyptus varieties, coconut, silk cotton and the neem tree. Almost all the trees listed at Appendix C are found in the vegetational zones mentioned above.

Beekeeping could be practiced all over West Africa. But commercial beekeeping must be restricted to the grasslands. The desert oasis and the deciduous forest may support small-scale beekeeping activities to promote adequate pollination of citrus and fruit crop farms to increase yields. Bees can also be kept in the equatorial forest and the mangrove swampy areas but the aim must be pollination rather than for honey and wax production. We therefore encourage beekeeping all over the land. There is no area that should be over-looked as even the desert has bees. All energies and resources must be used to harness and exploit this available natural resource. The inexperienced beekeeper is advised to follow the simple instructions given in this small book to reap maximum results.

Marketing of honey poses no problem in this region of sugar and food shortages. Concerning beeswax, the TCC and Mansusu Agencies of Kumasi, have begun purchasing the material with the intent of feeding industry and crafts. Senegal, Guinea Bissau and Nigeria are known to sell the stuff and this writer will be very grateful for any information on this subject so as to make the next edition of this book more complete. We have not started selling propolis. We may do so when beekeepers show an interest in collecting it. We will try to interest the government of Ghana and other foreign supporters to provide funds to train people to undertake the extraction of pollen, bee milk and bee-venom in the near future. Our aim is to exploit all available resources in the land.

To learn more about beekeeping, I am happy to inform readers here that funds have been made available for some experts at TCC to travel to other sister countries to assist them in establishing apiaries. This consultancy work can be done upon a request to the African Headquarters of the Food and Agricultural Organisation (FAO) of the United Nations.

Under this arrangement, individual bee enthusiasts may also travel to Kumasi's University of Science and Technology to study apiculture for a short period. The TCC organises short courses which usually last between 7 and 14 days. These courses are usually announced in the local Ghanaian papers and the *Ghana Bee News* at least two months in advance to enable foreigners to participate. The language used is English. All those interested in the extension of this service should first contact the Director of the African Headquarters of FAO, Box 1628, Accra, Ghana or the Director, TCC, UST, Kumasi, Ghana.

To enable us to offer the right leadership in the bee industry the TCC is a member of the International Beekeeping Association and has established strong connections with the International Bee Research Association and other world bee bodies. We also invite bee-experts from other countries to assist us during our beekeeping courses.

This small handbook endeavours to give you most of the information you need to make a start in beekeeping. So why not try now?

CHAPTER ONE

Equipment

The first thing one needs, to be called a beekeeper, is a beehive. This is made of any wood which has the following qualities. It should be:
  (i) warp-proof;
  (ii) resistant to the rotting effect of sun and rain;
  (iii) termite-proof.

The Beehive

The TCC beehive is made of *emire (terminalia ivorensis)*. Beehives made four years ago have shown no sign of deterioration and it is believed that the hives will last for more than 20 years in the sun and rain. Other woods which can be used are locally called *odum (chorophora excelsa)* and *dahoma (piptadeniastrum africanum)*.

Beehives are sold in Ghana at (a) TCC, UST and Kejetia, at Mansusu Agencies in Kumasi (b) Intermediate Technology Transfer Unit (ITTU), Tamale, and (c) The Association of People for Practical Life Education. P.0. Box 106, Atebubu. The modified Kenyan Top Bar Hives are sold as standard equipment.

There are many different types of hive, but we shall concern ourselves with only a few. Kenyan Top Bar Hives were the first hives used. now due to some local problems the Kenyan Top Bar Hive has been modified at the entrance (Figure 2).

The Kenyan Top Bar Hive is fitted with V-Section removable top bars (see Figure 1(v) and 1(c). The V-section forms a starter and there is no need to attach a comb foundation to show where the bees must build the combs. The grooved top bars (Figure

FIGURE 1: Beehives, top bars and frame.

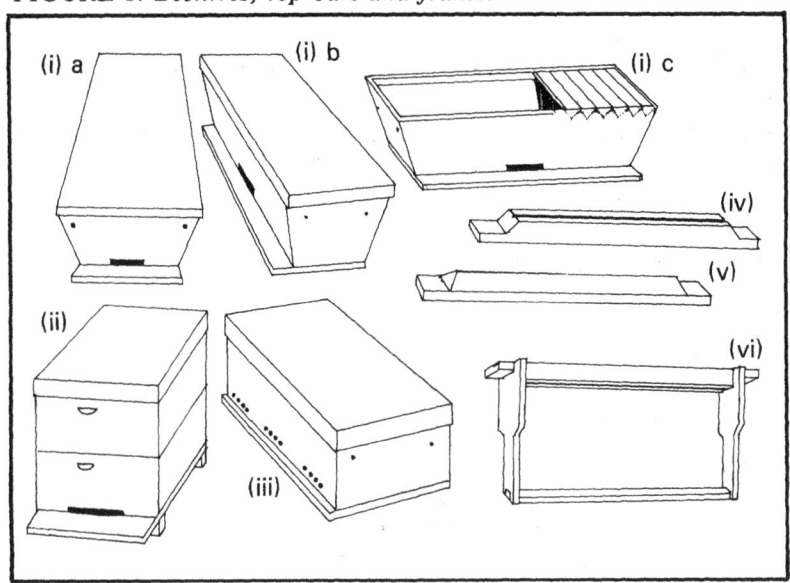

FIGURE 2:
(i) Kenyan Top Bar Hive, as used in Ghana.
(ii) A swarm catcher.

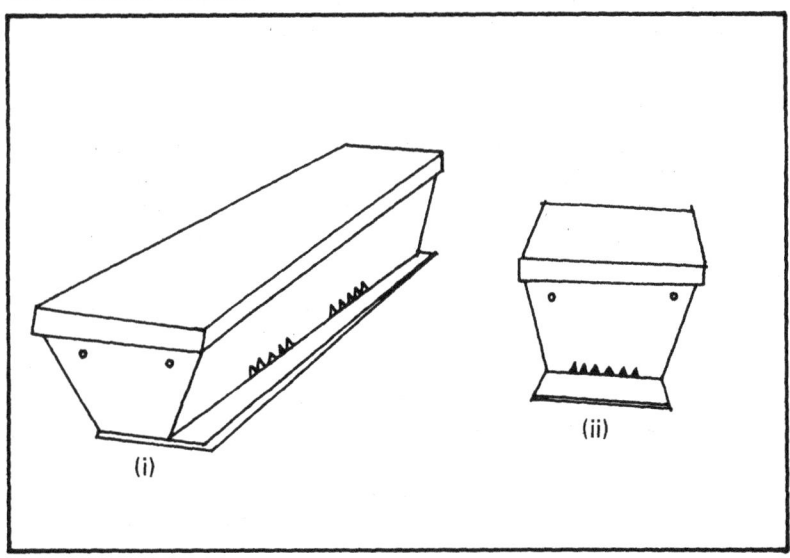

1(iv)) instead of the V-section could be used but it is strongly advised that a comb foundation of beeswax is needed to form a guideline. Otherwise the bees may not follow the line of the top bars but will cross them instead. This creates a serious problem which will give no room for inspection of individual combs. In order to avoid problems, it is best initially to purchase hives from one of the suppliers listed above.

Tanzanian Top Bar Hive

The next hive is the Tanzanian Top Bar Hive. This uses the same top bars as the Kenyan model. It has the same dimensions as the Kenyan Top Bar Hive but the shape of the base differs ((iii) of Figure 1). The shape of the Kenyan hive resembles a coffin whilst the Tanzanian is a rectangular box. The entrance could be like a slot or a set of tiny holes 8-10mm in diameter or small ʌ shape as Figure 2 illustrates. The ʌ shape became necessary when it was learned that such a shape is easier to construct than the round one and can protect the colony from attacks by beetles, large moths and any large creature which cannot enter through slots. With the introduction of this entrance, the Ghanaian modified Kenyan Top Bar Hive prevents the infestation by such moths as the *acherontia atropot* which previously posed difficult problems.

The Long Transitional Hive

The Tanzanian Long Transitional Hive assumes the shape of the rectangular hive described above but contains not top bars but frames (Figure 1(vi)). This type of hive resembles the European hive but it is not built in the same manner as the hive described below.

The last of the hives available is the Tanzanian commercial Beehive. This is a small beehive with frames and the same depth as the long transitional hive but contains between 10 and 13 frames only as compared to the long transitional hive which has 27 to 30 frames. This makes the commercial hive look more like a cube. Another apartment can be built on top of the base to become a second storey. Two and three storeys can be built and a queen excluder can be used to separate the brood chamber from the honey chamber.

For beginners, the TCC has chosen to employ the Kenyan Top Bar Hive for its simplicity. What we must know is that the industry is very new and with practice we will move from the simple to the more complicated. The TCC has all the types of hives mentioned here.

The top bar hive has been chosen because of the following advantages:
1. It can be opened easily and quickly.
2. The bees are guided into building parallel combs by following the line of the top bars.
3. The top bars are easily removable and this enables the beekeeper to work fast.
4. The top bars that hold combs can easily be determined so that opening of the box is started from the empty side.
5. The top bars are easier to construct than frames.
6. Top bars are cheaper than frames. At the time of writing one top bar costs 6 cedis whilst one frame costs over 20 cedis and moreover, it is difficult to find a carpenter who will construct the frames readily.
7. Bees in the top bar hive can easily be controlled when harvesting or while inspecting the combs. The smoker puffs smoke through the opening created by the removal of one top bar; few bees are then allowed to come to the attack. With the long transitional hive, when the top cover is opened all the frames are exposed which permits the entire colony to attack. Only a very powerful smoker will be able to control them. In practice, it is easier to control bees in the top bar hive than in one with frames.
8. Even with bees it is light and easy to transport.

Below are some disadvantages of the top bar hives:
1. From point No.8 listed above, it will be noted that its light weight and shape makes it easier to steal; however, the new industry has never yet had such a report. If any beekeeper is threatened by the danger of theft, then precautions should be made to either set up the apiary at a safe location or employ a watchman.
2. Combs suspended from the top bars are more apt to break off than those which are built within frames. This makes it difficult to transport a colonised hive in vehicles especially on

bad roads for long distances. V-shape top bars suffer from combs breaking easily and this is the reason why the grooved top bar was introduced but that does not solve the problem completely.
3. A top bar hive with provision for a lock has been found to be necessary since the Centre experienced cases of theft in which two colonies were burnt in Kumasi. Another instance occurred in Accra where a total of seven colonies were destroyed by fire and several kilograms of honey were stolen.

Top bar hives with locks are also produced at TCC on request.

After obtaining hives, the next items required are a smoker, a veil, a beesuit, a knife or a hive tool, a pair of gloves, a pair of shoes and a brush or quill. Others are a swarm catcher and sometimes a match box.

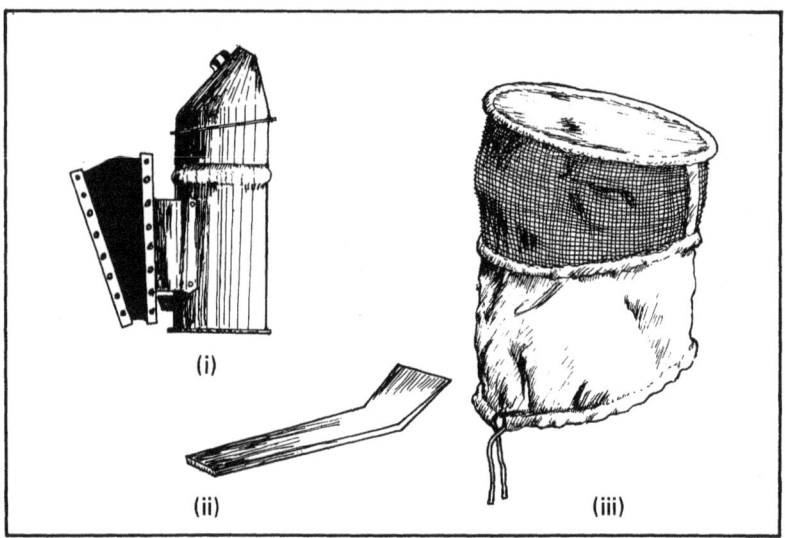

FIGURE 3: *The smoker (i), hive tool (ii) and bee veil (iii).*

The Smoker

This is the next in importance after a beehive. The tropical honey bee will not allow the beekeeper to take away honey with ease. Therefore a good smoker which is capable of puffing sufficient smoke to render the colony docile must be acquired: Figure 3(i). The TCC usually has large quantities of smokers for sale to beekeepers.

The Veil

A veil like that shown above can protect your face from bee stings. Be sure to put it on very well because if any bee succeeds in entering, it will be very difficult to work. Use the twine hanging below to tighten the veil around the neck firmly and make a knot around the waist to secure the veil.

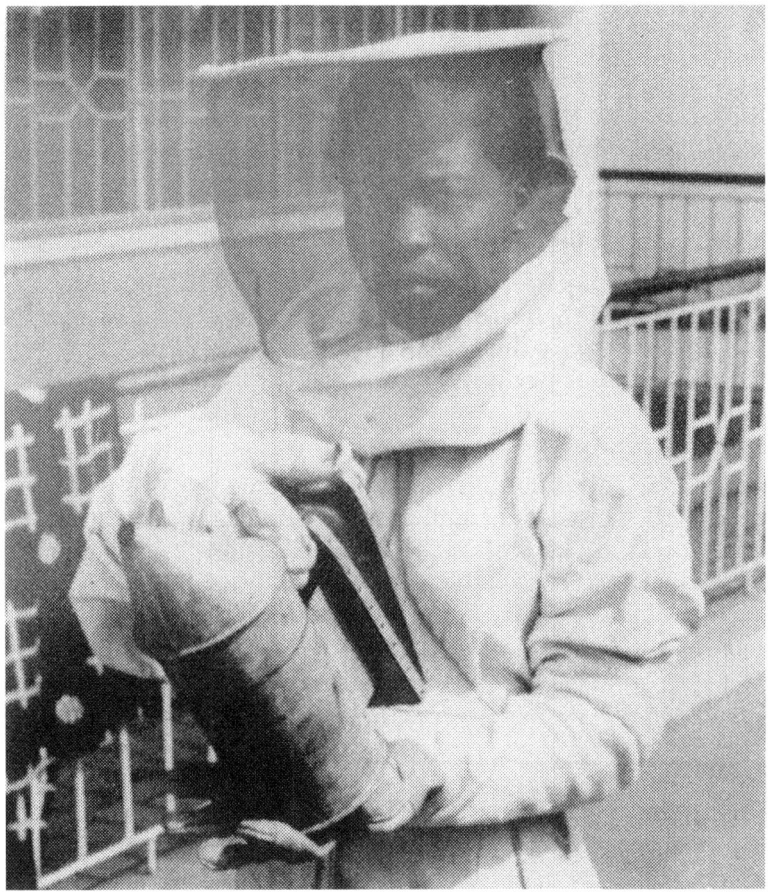

PLATE 1: A beekeeper in a beesuit, veil and holding a smoker.

The Beesuits, Gloves and Shoes

These are all important and must be acquired in advance before the honey is harvested or any work involving the opening of the hive is undertaken. The beesuit is sewn in a way that the whole body is protected except the head, hands and feet. Beginning beekeepers need beesuits, a pair of leather gloves and a pair of 'wellington boots' to protect the hands and feet. As the beekeeper becomes experienced he may begin to discard these protective clothes one by one.

A Hive Tool and a Knife

A hive tool or knife is used to detect the empty side of the hive and to remove the top bars. In fact it is more necessary to acquire a knife which has an additional duty of cutting the honey combs. A knife can perform all the functions of the hivetool but the hivetool cannot be used to cut the honey comb neatly.

A Brush or Quill

One of these may be used to sweep the bees from their combs. If a brush is used then the tips must be soft enough to protect the bees and not injure them. The quill or feather from a large bird like the turkey is preferred to a brush.

A Swarm Catcher

Figure 2(ii) This is a small beehive containing five or six top bars or frames. As its name implies, it is used in securing a swarm of honey bees for the purpose of hiving. It is installed up in a tree so that when it attracts bees they are emptied in a careful and orderly manner into the beehive which is four to six times bigger than the catcher box. A swarm catcher's top bars or frames have the same dimensions as that of the beehive, this makes it easy to transfer bees and bee combs from one into the other.

A Match Box

This is needed as an apartment for the queen bee when a swarm or a colony is being transported. This ensures that the queen is always safe throughout the journey and she does not become lost.

CHAPTER TWO

The Colony

A colony of honey bees, the *apis mellifera adansonii,* is composed of the following:
(a) One queen whose main work is to lay eggs;
(b) about 300 to 500 male consorts of the queen, who are called the drones;
(c) 6,000 eggs, to be hatched in three days;
(d) 10,000 to 15,000 larvae which need to be fed and kept comfortable by clustering workers. A temperature of 35° centrigrade must be maintained to keep them warm.
(e) about 15,000 to 20,000 sealed brood which need no attention except continued warmth;
(f) About 20,000 to 25,000 home or domestic bees whose duties are to clean the hive, build combs, feed the brood and protect the colony from its enemies;
(g) and about 20,000 to 25,000 foragers who collect pollen, nectar, water, propolis and everything that needs to be brought into the beehive.

Let us now have a look at the honey bee who is the hero of our beekeeping industry. There are two sexes, namely the male and the female. The male is known as the 'drone'. The female exists in two types; the 'queen' and the 'worker'.

The Drone

The Drone is the laziest of the three and his presence in the hive seems to be of less importance to the beekeeper. He is stout and larger than the worker. He has no suitable proboscis for gathering nectar and has no sting to defend himself or the colony. He possesses no baskets for collecting pollen grains and no glands to secrete wax for building combs. He does no work in the colony but eats and moves about making loud noises everywhere he goes. He is therefore considered useless but has a very important function which only a few of his kind ever fulfil. This function is to impregnate the queen, which is done only once, the drone dying immediately afterwards. This official duty, the mating with the queen, can be done ten days after emergence as an adult. Drones only go out of the nest at midday when the weather is fine and warm. They may live for twelve or sixteen weeks.

PLATE 2: (i) Drone (ii) Worker (ii) Queen (by courtesy of Dadant & Sons Inc.)

The Queen

The queen, of which there is only one in the hive, is the longest of the three. She is larger than a worker and longer than a drone. Her wings are much shorter than her body. The wings cannot clothe the whole of her abdomen. Her long tapering abdomen makes her appear like a wasp. She has sparkling golden hairs on her skin and like the drone she has no sting. She has no proboscis and no basket for collecting pollen. She does not secrete wax. She does not go to the fields to collect nectar or feed herself. She goes out once in her lifetime for the purpose of mating. She is mentally less developed than a worker bee. She is like a slave because she does not go out to play or anything. She goes out only when the whole colony decides to abscond to another hive elsewhere; a phenomenon which does not exist among other bees in Europe, only in the tropics and in the East.

The queen's main duty is to lay eggs; and this she does — laying between 1,500 and 3,000 eggs in a day during the peak of the brood rearing season. She is described as an automatic egg laying machine. Unfertilized eggs laid in the hive hatch into drones whilst the queen's fertilised eggs produce workers.

The Worker

The worker is the smallest and most numerous of the bees and constitutes about 98 per cent of the colony's population. One colony may boast of 60,000 to 80,000 workers.

One may call a worker bee a virgin or a nun for she is never mated. She lays eggs only at 'emergency' periods when the queen dies and she would like to function as the queen in order to maintain continuous life in the city. However, this effort always ends in futility. The eggs laid by a worker are unfertilised and only hatch into drones.

She has the right type of proboscis to suck juices, nectar and water. She collects pollen grains and puts them into her two baskets and transports food for the whole colony. Without her, the colony is doomed and there would be no honey and wax. She collects nectar and changes it to honey. She secretes wax to build comb cells which act as cups to contain honey and 'rooms' or 'chambers' for broodrearing. She collects resinous material from trees which is called propolis. This substance is used to block holes and cracks, to repair combs, to strengthen the thin borders of the comb, and for making the entrance of the hive water-tight or easier to defend. It is used to cover objectionable things; it is used as an embalming substance to embalm dead intruders like the wax-moth, snakes etc. which are too large to be moved from the hive.

The worker bee transports into the hive and out of the hive

anything that needs to be transported. Any heavy material or dead body that cannot be transported easily is dragged along and thrown overboard. The worker bee is always busy. She works tirelessly especially during bright sunny days. The worker bee is a product of a fertilized egg from the queen.

The queen lays an egg and a nurse bee places it in the midrib section of a worker cell. The egg hatches into a larva and in three days. All larvae are mass fed on royal jelly (a special food produced by the workers and used to feed the queen) for the first two days. On the third day of the larval stage, only a queen-to-be larva is continuously fed with the royal jelly which is sometimes called bee-milk. All worker-to-be larvae are fed on a poorer diet, that is, honey mixed with pollen and diluted with water. While the queen-to-be is given more food than she can eat all alone, the worker-to-be larva is 'starved' a little so that she does not grow as big as the queen. It takes an egg 15 days to become a queen whilst it takes 21 days to produce an adult worker bee. The pupa of a worker takes 12 days to emerge as an adult bee.

The newly emerged worker bee is very well 'educated'. Make no mistake in underrating her for she knows everything and understands whatever every bee has to know. She is programmed by nature. The only thing she has to learn is the location of the hive and the geography of her surrounding land.

The kind of work performed by the worker depends largely upon her age. The first three weeks of her adult life (or six weeks during the honey production season) are devoted to activities within the hive whilst the remainder are devoted to field work. Among all the various duties, comb-building and brood-rearing may be considered the major activities. Comb building provides shelter for reproduction — in fulfilment of fundamental biological needs.

Wax and all material used by the honey bee to manufacture combs are produced of her own body. These are secreted through the glands of the worker at the height of her development; that is 12 to 18 days old. The wax is secreted through and between the overlapped portions of the last four abdominal segments on the underside of her body. She converts 15kg of honey into 1kg of wax. Wax can be secreted only at relatively high temperatures from 33 to 36 degrees Celsius or 92°

to 97°F.
Other duties of a house bee in the hive are:
  (i) cleaning operations
  (ii) feeding or nursing the brood
  (iii) comb building
  (iv) administering to the queen
  (v) orientation flights
  (vi) ventilating the hive
  (vii) guard duty
  (viii) pollen, water, nectar or honey packing and
  (ix) executions

Cleaning Operations

This is the first activity of a worker bee. She begins to clean and groom herself immediately after she crawls out of her cradle. When she completes removing unnecessary particles and dirt from herself, she takes in food and then starts cleaning the brood cells for the queen to lay her eggs in. The queen, before laying an egg into a cell, examines it to satisfy herself whether it has been properly cleaned. If she finds that the cell is not properly cleaned she quickly rejects it. When a young worker is executing this cleaning operation, she uses both the tongue and mandibles. After cleaning continuously for three or four days the young bee takes on the duties of a nursing mother. The cleaning activity will cease but may resume as occasion demands. Other duties which may occasionally be necessary are (a) removing from the hive dead intruders or dead bees and (b) removing debris and other objectionable objects. Anything that is too large to carry is often dragged along and pushed overboard. Any dead snake or waxmoth, or any carcass that is too heavy to transport, is embalmed with propolis. 'Cleanliness is next to godliness' and the worker bee seems to demonstrate this proverb.

Ministering to the Queen

The next work on the programme of a young bee undertaken after the third day of emergence is to provide for the needs of the queen. At any time the queen needs food, she asks by stretching out her probiscis towards the mandible or mouth of the nearest worker. The workers are always anxious to satisfy her needs; wherever her majesty is, the attendants make a circle or a semi-circle around her. Every nurse bee in attendance watches closely and is very anxious to be the first to answer the request. However, not every worker or attendant is able to provide the need at every instance. The queen contacts the nearest worker and if she does not get what she needs, she approaches the next. This goes on until all her demands are met. It is the duty of the attendants to bathe her. This is done with their tongues and mandibles. Her faeces also have to be carried away by the nursing bees.

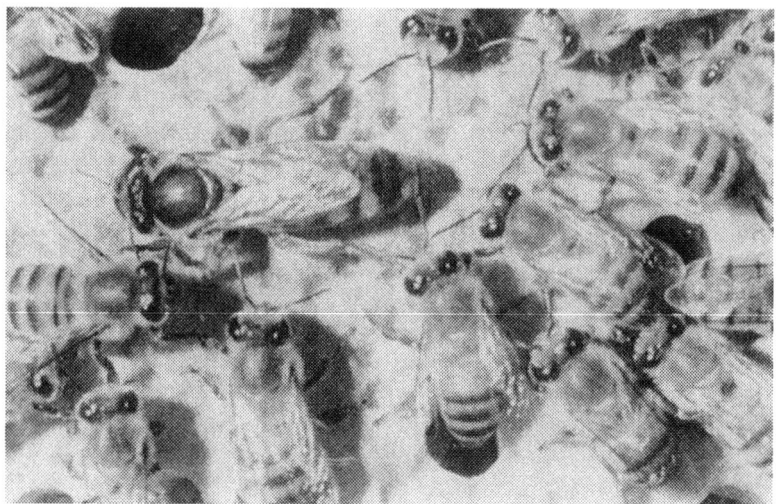

PLATE 3: *Nursing bees in a circle around the queen (with permission of Mr. Bernhard Clauss).*

Orientation Flight

This is the next exercise for the young bee. She tries some short flights in front of the hive and in its immediate vicinity. This enables her to acquaint herself with the environment so that when in the near future she goes out to forage, she will be able to find her way back.

Ventilating the Hive

Bees try to control the temperature of the hive. When the

weather is too warm for them, worker bees fan with their wings in an attempt to reduce the temperature. They also have to fan uncapped honey combs to remove any excessive water from them. Water is also to be brought for this cooling business. They cluster together to keep warm when the weather is cold.

Guard Duty

This is one of the last duties for a house bee before she engages in field or foraging business. By this stage, the house bee has attained the apogee or the height of her youth. She is very energetic and is suited to defend the city. This work is executed with devotion and a keen sense of patriotism. It is done without any obligation. The bee soldier is ready to die for the benefit of her majesty and the entire colony at any moment. Any intruder, robber, or any enemy would receive first a frightening verbal warning followed by a sting and if it persists an alarm would quickly summon more defenders to face the foe.

Executions

This function is deemed necessary and performed as a means to protect the colony from hunger and robbery. Executions are performed for three main reasons:
    (i) to eliminate useless drones at the time of hunger
    (ii) to kill unemerged brood and
    (iii) to eliminate any robber bee or any enemy.

Execution of Drones

This sad event mostly occurs after the queen's nuptial flight and when the queen has successfully started laying fertilized eggs to expand the city's population. By this time, there is little food in the hive but the drones who have just completed their mission (i.e. impregnating the queen) continue to be fed on the little honey left in the cells. To reduce this drain on scarce resources an alarm is sounded and the workers execute the 300 to 500 drones in the colony. Against the attack of the workers the drones have no defence. They are literally torn to pieces amidst a characteristic wailing that the beekeeper comes to recognize.

After a couple of minutes, there is dead silence followed by a mass dragging of drone corpses to be thrown overboard. A very few lucky drones may survive this massacre. Those that remain in the hive may live for a total of three or four months before their natural death.

Robber Bees

All worker or foraging bees are thieves. They claim anything they like as their own property. They would visit frequently any sweet juice or any water found elsewhere. Robber bees try to visit another colony's hive and try to take honey with the view to storing it in their own hives. But thanks to the good work of the patriotic soldiers at the security gate, they would not be allowed to enter without a fight. It is strange that if you put water or any juice very close to the hive, the bees seem not to take advantage but if it is placed further away, about twenty yards or more, then they take it. This illustrates to beekeepers that they should always watch their hives well to avoid leakages of honey. When this happens, the bees in the hive will not take advantage to collect it back into the combs. Leakages will only attract robber bees from another hive or wild colony.

The Field Bees

Work involving flights begins as early as the third or fourth day but not many trips are made at first. These flights, as we have seen, are done to acquaint the bee with her immediate surroundings so that when foraging business begins later she will be able to return to her home safely.

A 'fielder' normally starts pure field work at the age of three weeks (or thereabouts). Work performed may be fetching water, nectar, pollen and propolis. Sometimes one bee may carry two of these items i.e. nectar and pollen in one trip. Unlike butterflies and other insects which visit a flower and take the nectar only when they are hungry, bees collect nectar or juices and fly quickly into the hive and discharge the load. This they usually transfer to the house-bees for packing and safe keeping. Then they quickly fly out for the next trip. The field bee tries to go on as many trips as possible provided there is more afield to harvest. There is no time wasted. She flies at an average speed of about 50km per hour though this is dependent on the weight of the load being carried.

Impregnating the Queen

This is termed the nuptial flight. It is done five days after the queen's emergence from her brood cell. Immediately after she emerges from her cell, the queen tours every corner of the hive to see if there is another rival queen. If one is found, there will be a fight until one overpowers the other and kills her. If the colony has decided to swarm, then the house bees will prevent the newly emerged queen from fighting with the old one but she will still go round with the sole aim of destroying queen cells and their contents.

The young queen may join a swarm and leave the parent hive for the onward march in search of another dwelling place. When a suitable abode is found, the next important job which must be done and which is the bedrock of the empire, is the impregnation of the queen.

The new unfertilized queen has as many as 500 potential husbands but does not need all of them to mate her. Four, five or seven are enough. She would choose a good bright day devoid of thunderstorm, lightning, fog, cloud, wind and rain or shower. Nature has for her many enemies. These are birds, reptiles, spiders' webs and other dangerous insects like the bee pirate. Any of these natural hazards could contribute to the death of the young queen which could spell the doom of the entire colony.

The queen flies out accompanied by most of the drones. She flies very high and there one of the 500 succeeds in mating with her. The successful drone then dies as its male organ is broken and sticks into the vital organ of the queen. The rest return with her majesty into the hive again. One of the workers would help to remove the broken male organ from the queen's body and it is quickly thrown away. The nuptial flight may be repeated several time that day if the fine weather continues.

Sometimes, nuptial flights can be delayed. This may be the result of a long rainy season or pronounced bad weather. When this happens,[1] a young queen bee may no longer have this very important natural phenomenon accomplished. In this case, the

---

1. The TCC experienced this phenomenon in July-September, 1980. The old doomed colony was replaced by another strong colony which entered on 1st September and quickly occupied 8 top bars with combs.

PLATE 4: *The mating queen bee (by permission of Dr. Eva Crane, IBRA).*

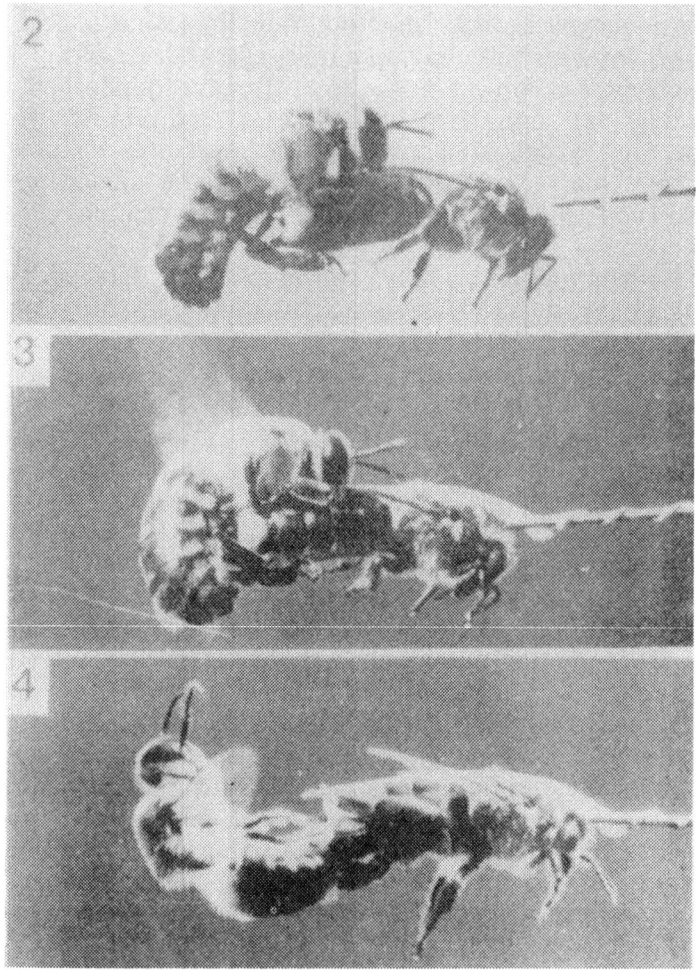

Fig. 2. The drone seized the queen, mainly with his first and second pairs of legs.

Fig. 3. The drone succeeded in holding the queen ⅓ second later, with all his legs in the final position. He is shown lifting the posterior end of his abdomen to contact the open sting chamber of the queen.

Fig. 4. Eversion took place 1 second after Fig. 3, and the paralysed drone then swung away backwards and downwards.

colony will perish within a few weeks unless the beekeeper receives the message and reacts immediately by inserting an egg comb from another hive.

The Unfertilized Queen

The unfertilised queen will lay eggs but these will hatch into drones. After four to five weeks, all workers will have perished and drones will remain. They, as you have been told, do not fetch water, nectar or pollen and indeed, they do no work for the colony. Therefore, there will be an acute shortage of essential commodities and services. The queen and all the drones will later perish of hunger and thirst because they have no worker bees to serve them.

This is one of the reasons why you have got to inspect your hives regularly to effect proper control, especially after it has been newly colonised.

Table 1: Development of brood in days

	Queen	*Drone*	*Worker*
Egg period	3	3	3
Total larva feeding period	6	6	6
Cell canned pupa	6-7	15	12
Total Period	15-16	24	21

The Queenless Colony

It is very interesting to study the behaviour of a queenless colony. A queen-right colony has all the workers performing their normal duties i.e. carting pollen, nectar, propolis, water, all the sorts of work that have been described above. The queenless colony can be likened to a country or a human society without an accepted ruler; a state of anarchy is in effect.

When a queen dies, the workers find a young worker-to-be larva which is less than two days old, or a female egg. The bees collectively agree to build a queen cell around this 'constitutionally' selected larva. To make sure they get a queen, two or more queen cells may be built around such elected larvae. They feed them with bee milk or royal jelly until the first to emerge takes over or supersedes the old queen. The workers will then help the emerged queen to kill the unemerged ones.

In a colony where such potential queens-to-be are not available, some of the larger workers begin to develop ovaries and make themselves queens. Every large worker who makes herself a false queen tries to solicit support from the smaller ones. Thus several false queens begin to reign the city. 'Power' is now coming from several false queens. The bees become divided and there is no clear line of policy to follow. When they decide to abscond, they are divided into as many groups as there are false queens. If they remain in the beehive or the old nest then the false queens begin to lay eggs. One laying worker places her egg in one comb-cell and another laying worker lays her egg in the same comb-cell so that two or more eggs are crammed into one cell. This pandemonial state of affairs does not end here. Worker bees, upon learning that they have no 'popularly' accepted queen to leave them will refuse to accumulate nectar in the comb cells but keep the nectar to themselves. Pollen gatherers stop work. No bee will build combs or cart propolis; in effect the normal business in the city comes to a standstill.

Each worker begins to collect for his needs and this results in a state of 'power to the workers'. No pollen comes into the hive as there is no brood to feed. A worker bee in such a colony may live for more than four months. Remember that such a colony can never rear a new queen even if you supply a set of brood combs. This is because their glands that secrete royal jelly have become inactive, so the workers would not be able to feed the young larvae. In this case the best thing to do is to add the remaining bees to an existing colony. (Refer to uniting bees).

CHAPTER THREE

Hiving a Colony

Once you have acquired your hive, the next thing you have to consider is how you will get a colony or a swarm of bees to stay in (colonise) the hive. Provided it is the swarming season colonisation does not pose any problems in West Africa, a region which is full of wild colonies. The swarming period differs in various parts of West Africa. In Ghana, in the forest around Kumasi, it occurs between June and October whilst at Ejura, Nkoransa and Atebubu, in the transition forest zone, it occurs between November and February. As a beekeeper you have got to study when the swarming takes place in your locality. During this period, old colonies divide themselves into smaller groups and they fly away and never come back to their former dwelling place. Now let us study what happens during the swarming period.

The Swarming Season

In the last chapter, we learned about how the queen reproduces individual bees. The reproduction of an individual worker, drone or queen is distinguished from the reproduction of colonies. The honey bee colony has been endowed with an instinct which brings about an increase in the number of colonies from time to time. One colony may produce more than 10 colonies or swarms within one calendar year. This is the result of a plan made by nature so that when a colony in a nest or hive is too populous, the old queen, accompanied by about 500 drones and thousands of young and old workers fly to a distant place to begin life anew. No single bee or group of these new settlers will ever return to the old nest with the view of paying a visit or going to fetch their property which was accidentally left behind. The old hive is forever forgotten. As the bees leave the entrance of the old hive they fly gyratingly into the sky with a loud hum. They will soon go and land in a branch of a tree to cluster. We refer to this cluster as a swarm of bees.

The swarm hangs there temporarily. That is neither the end of the journey nor where they are going to stay and build a new city. They send some scouts to go and find a hollow tree or any suitable place for the new empire. This place could be your hive. The exploratory team of scouts, if lucky, will return with a favourable report to her majesty and her subjects still waiting on a branch of a tree. The swarm will now leave following the scouts into the new-found home.

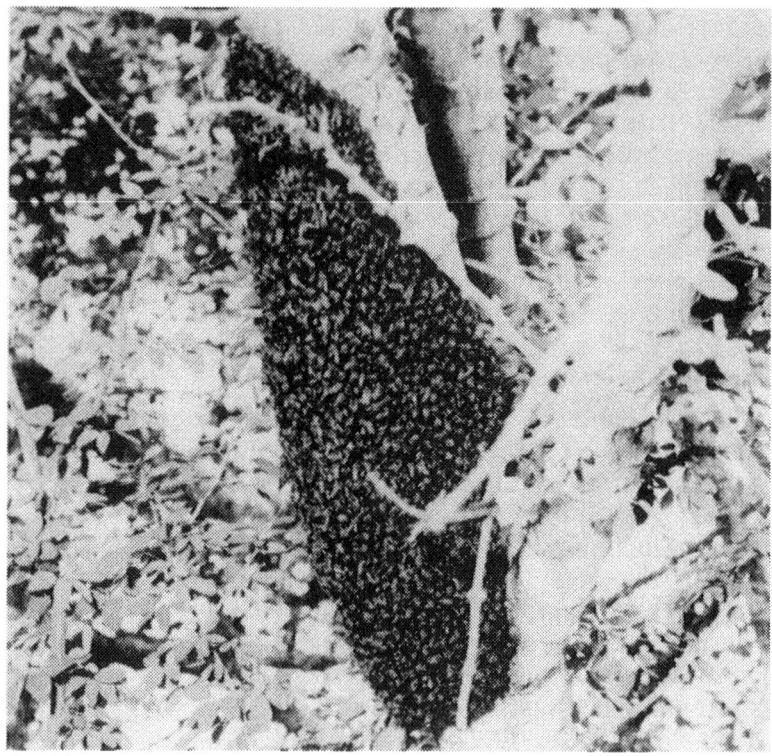

PLATE 5: A swarm.

In every year, the first swarm that leave a hive is called a prime swarm. A prime swarm is always accompanied by the old queen and some older workers. Before leaving the old hive, they take in honey and other essential commodities from the old nest. This is so that when they settle in the new nest, they can begin to build combs within a short time. The queen does not need to be impregnated and soon begins to lay eggs to build a new strong colony.

After a prime swarm, any other batch that leaves the 'parent' hive is termed a secondary swarm. All the batch is composed of young workers, young drones and a young queen. They are completely docile as they are seen hanging in a tree or when they colonise your hive. They may begin to show some aggressive tendencies after six or seven weeks. They are very slow to work. The queen will have to go for her nuptial flight for impregnation within five days and egg laying will start on the eighth day. These young bees may need you to provide sugar syrup as a supplementary food to help them for some time. If you are unable to feed them, they will survive but your assistance may enable them to work faster than if they had received no help from you.

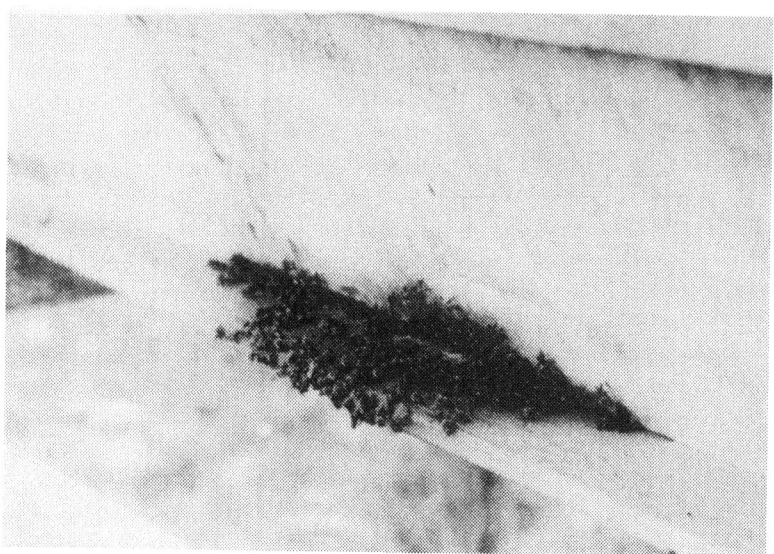

PLATE 6a: Slot entrance.

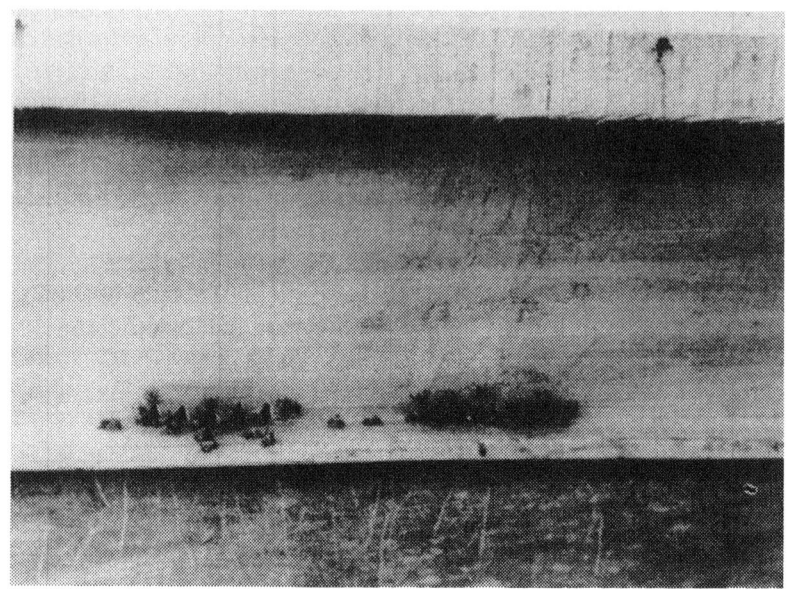

PLATE 6b: Triangular-notched entrance.

The Entrance of the Hive

By nature, the honey bees' most important task after entering a new hive is to redesign the entrance with propolis to conform with their own taste. Within the first week, more time is wasted in collecting propolis from buds or trees and used to close up the gap made for them to use as a gateway into the hive. By so doing, birds, reptiles and bigger insects like beetles and butterflies are kept away and cannot enter. The remodelling keeps most natural enemies away and gives greater protection to the weak colony. It also prevents water from entering from the outside platform even if the hive is tilted inwards. (Several times I have watched a hive which is about to be colonised and made this observation which I have not found recorded in any beekeeping book. Some of 'the scouts' may choose to stay in the hive and clean it. They will remove dirt and broken pieces of wood and throw it overboard. This house clearing exercise may continue for several days before the arrival of the bee contingent. In most cases, however, it takes less than an hour for the preparation to be completed and the swarm to arrive. The record time for colonization in the Kumasi area was 20 minutes.

During my reading of the literature I have noted the following and have confirmed it by my own experience. In an apiary where there are several empty hives an observer might think that more than one hive will be colonised. This is due to the fact that, on arrival of the bees, a number of workers and drones will leave the swarm and enter several beehives. However, when these bees do not find the queen within the respective hive, they will begin to search for her. When she is found in one hive, they will then join together again to form one colony.)

How to Capture a Swarm

Do not wait for a swarm to colonise your hive if there are any swarms hanging in your vicinity. All you have to do is to capture them for your hive. Now here are some useful hints to help you.
1. You may fill your smoker with dried wood shavings or any dry material that can provide you with sustained smoke throughout the operation. (A courageous beekeeper needs no smoker for catching a swarm).
2. You may also put on your beesuit and veil.
3. Get your hive, catcher box or any clean odourless container. You may use one of your veils (the TCC Veil) which can serve as a bag to capture the swarm.
4. Sprinkle water on the swarm. You may use a sprayer if you have one.
5. Climb the tree or to wherever the swarm is hanging, and shake all the bees into your container. Wait for ten to twenty minutes. If the queen is in the box you will know as all other bees begin to enter your box. If you want to make sure you have got the queen, catch her and place her in a match-box. Open the match-box slightly and pin it at the upper end of the catcher box. This will enable bees to cluster around her comfortably.

You may by-pass some of these steps by using a knife to cut the branch off the tree that contains the swarm and carry it home for hiving. Catching a swarm must usually be done late in the evening to ensure that most foraging bees are back to base to join the cluster.

Hiving the Bees

It is advisable if you have an old hive with a colony of bees to insert a brood comb from the old hive into the new hive. Now attach the queen to the brood comb you have placed in the centre of your hive. Leave all bees behind in the old hive. Bees from the old hive cannot live with the new bees. Only drones can be accepted in any hive if there is sufficient food in the host hive. You may now shake the bees from the captured swarm into the hive. With the old brood comb the bees will feel 'at home' and therefore accept the hive readily. Allow time for all the bees to settle before dressing the hive with the remaining top bars. Cover with the top cover. Seal the entrance and do not allow the bees out for 24 hours. Remember to provide food (sugar syrup or honey) if possible.

If your top cover is metallic, place a thick bundle of dry grass or any dry leaves on top of the cover. This is needed as a shield to avoid overheating in the hive. You may now hang the hive in a tree or erect a platform with bricks or wooden legs. Use grease to coat the underneath part of the platform's legs or blocks or on the cords or the string which has been used for hanging the hive. The purpose of the grease is to keep all ants away from the hive as they cannot creep through the sticky grease.

There is another interesting way to hive a swarm if they are hanging where you can easily reach them:

1. Arrange all the top bars neatly as they should be but leaving out one or two.
2. If you can secure a brood-comb from an old hive then you may insert one or more (with no bees from the old hive). You may ignore this section if you cannot secure a brood comb and go on to number 3 below.
3. Remove one or two top bars and with a white paper in one hand, and a big feather or a soft brush in the other hand, brush off a small cluster of bees from the big cluster onto the piece of paper and place the bees into the hive.
4. Now cover it with (i) the last one or two top bars and (ii) the top cover to the beehive.

    You now still have the big cluster in the tree and it is now time to finish up and you must act quickly so that the bees you put into the hive will not join their colleagues in the tree.
5. With the help of the brush or quill collect more bees from the

PLATE 7: Bees for the hive. Never be afraid of a swarm of bees. Learn to handle them coolly like this.

cluster onto the white piece of paper and this time, place the bees (on the paper) at the entrance of the hive. The bees will rush quickly to join those already in the hive. Repeat this process until all the bees have gone into the hive.

Caution
If at any moment during this operation the bees decide to leave, please act quickly to block the entrance, preferably with beeswax. If there is no beeswax available then a piece of wool or a plug of grass can be used. The bees will then begin to chew the wax in order to remove the plug. By the time the bees have finished, they may be calm again and will stay in the hive quietly. It is then possible for the beekeeper to continue the exercise by placing the rest of the clustered bees at the entrance.

Note
Do not fear to do this; the bees are not so dangerous at this time. If you are stung remove the sting and the scent sprayed on the spot of the sting. Check to be sure that the bee smell is not there before continuing the job.

(The writer employs school children from the age of 5 to 10 to do this operation. The children are very happy to perform this exercise. The photograph of the writer with a swarm of bees shows how a swarm can be handled. You can do the same).

6. The hive may now be carried away and placed on a platform which has been raised previously. *Please do not hang the hive as the bees will be greatly disturbed during the hanging period if the hive is jarred.*

Preparing the Hive

Before hanging takes place, the hive must be prepared in order to attract the colonizing bees. To bait a swarm of bees into the hive, you will need one of the following items below:
a. A small cake of beeswax.
b. *Konkonte* (dry cassava) powder.
c. Granulated sugar.
d. Lavender with sweet smell.
e. A local herb called *nunum*.

f. A ball of lime.
g. Cow-dung.
h. Honey.[1]

Before attempting to bait, thoroughly clean any dirt from the hive.

(a) With the cake of bees-wax you must coat all or some of the ridges or grooves of the top bars as well as any rough surface in the hive body including the entrance. You may use a wax-dripper later on when you are treating a large number of top bars. In the absence of a wax-dripper use your hand to do the coating. Another thing you may use in the future is a comb foundation. This is a prepared sheet of beeswax which is used to bait a colony. This sheet of beeswax is never lost or consumed and may later be removed from the hive and melted in readiness for sale in the market.

(b) *Konkonte* (dry cassava powder) and granulated sugar are also used to attract bees. They are placed in the hive. Sugar is very sweet and is taken readily by bees to use for honey making. Dry cassava powder is used in the same way as pollen to feed young larvae.

(c) Lavender, *nunum* or lemon-grass leaves may also be applied in the same way by rubbing into the hive body so that the powerful sweet scent will attract the scouting bees.

(d) A ball of lime is placed near the entrance or on the top of the beehive. Cow-dung is placed in the hive.

(e) Honey may be the last resort if any of those baits mentioned above cannot be obtained. Like beeswax, it is a product of the bee and can attract hundreds of bees within some few hours. But the bees will take away all of it, leaving not a scent behind. In this way, there will not be any honey inside the hive to attract more bees unless the supply is renewed. Man cannot waste delicious food in this way for robber bees to load away so it is usually coated *on top of the top bars*. In this way some few drops will drip on to the base board of the hive to attract the bees. When this honey is finished,

1. Honey should be used as a last resort for the reason that some honey may be a product of a sick colony which could be dangerous to the in-coming swarm. You may not have to use it at all: after all there are several things available for baiting. The writer uses beeswax to attract all his colonies. It usually takes one or two weeks to attract a colony but this is always during the swarming season.

some will be left between the top bars and the lid which can never be reached. This will continue to give a powerful smell all the time so that one day, luck will 'catch' the 'scouts' that have been sent to find a suitable accommodation for her majesty and her subjects.

Forcing Bees Out

Another method you may apply to get bees for your hive (if your attempts to catch a swarm or bait the bees have failed) is to force or smoke wild bees out of their nest and hive them. This work must be done late in the evening.
1. Fill your smoker with fuel and put on your protective clothes.
2. Get your machete, chisel and mallet.
3. Gently puff some smoke at the entrance of the bees' nest. The bees will rush into the nest and gorge themselves full of honey. They will then feel too heavy and drowsy to move.
4. Use the machete, chisel and mallet as required in the operation. Puff some smoke at the bees any time they show any sign of aggressiveness.
5. Chop open the hive and begin to remove brood-combs.
6. Use a string, preferably, raffia string or a sinew, and tie the brood-combs to the top bars.
7. Capture the queen and attach her to one of the brood-combs in your hive.
8. Scoop as many bees as possible into the hive. Put in some honey-combs so that they may feed on them.
9. You may leave the hive for some few minutes to allow the other bees to trace their queen.
10. You may use a queen excluder to imprison the queen so that if the other bees decide to cluster nearby, the queen will not be able to follow them and later they will come back to settle in the hive.
11. Visit the hive at least twice a week to see whether the bees have clustered on the fixed combs. If they make no attempt to leave the hive you may remove the queen excluder from the hive. They have decided to live in the hive if you see that they are building combs.
12. Inspect it after one month to see whether the queen is laying eggs to increase the population of the community.

Hanging the Hive

The first thing you have to do when going to hang a hive in a tree is to inspect the tree to be sure that there are no ants in it. All types of ants are dangerous to bees. If there are ants, please avoid the tree. As trees are abundant in all bee areas, it should not be too difficult to select an appropriate tree.

Always place the hive underneath a shady tree. You may hang it in the manner seen in Figure 4 always bearing in mind to let the entrance or the occupied part of the hive face East, the direction of the rising sun. This will prompt the bees to start work very early. Do not forget to coat some grease on the cords. Renew the grease every two weeks until you are sure that your young colony is strong enough to repel any ant attack. Be sure you have hung it in a way that will allow you to easily work with it anytime the need arises. Tilt the hive so that the entrance is angled downwards to allow drops of water to trickle down through the entrance and across the platform.

How to Feed Bees[1]

Let us first ask ourselves what do the bees eat? The answer is nectar, sweet juices and pollen. In addition to these, bees need water for:
 (i) drinking and
(ii) for cooling the hive through evaporation during a warm day.

The best sites to place your hive are in orchards, on the banks of rivers or where the bees have easy access to water. The citrus farmer gains a lot if he hangs hives in his orange orchard for two vital reasons, namely,

 (i) the bees will pollinate the blossoms which helps the fruits set.
(ii) the bees, making use of the nectar and juices for which the farmer has no direct use whatsoever, convert these to honey and beeswax which he will later collect and use.

---

1. Ghanaian beekeepers now use yam-balls with cassava powder to feed bees. The yam is cooked and mashed smoothly and the cassava powder is added. Balls are now made and these are fed to the starving colony during the dearth season; or to a young nucleus. Place the yamball into the beehive.

So you see, the farmer's best friend is the honey bee. The honey bees acting as pollinating agents alone are enough to justify spending money in buying bee equipment. Indeed, honey bees are 'the angels of agriculture'. In Europe and other countries, beekeepers are paid for hanging their bees in apple and other fruit farms. Without the pollination services of bees, most fruits will never set.

You have just read that your new colony needs to be fed. This type of feeding should be provided directly in the form of sugar solution for them. You need a container like a jam jar or any bigger container. 500 grams of sugar should be dissolved in half a litre of water. Put the jam jar containing the syrup upside down in a shallow dish. Place a matchstick or something similar between the lid and the jar. Enough sugar water will be allowed to trickle out for the bees to take advantage of it.

With water, you may place a sponge into a container full of the liquid. The water will be absorbed by the sponge. The bees upon reaching the container, will land on the sponge, dip their proboscis into it and collect water. The bees are safe from falling into the water and drowning when this method is used.

If you cannot get a jam jar or sponge, you may collect water or syrup in any open container. Put in twigs or dry reeds. The bees will land on the twigs to take the food or water prepared for them.

FIGURE 4: How to install a hive.

CHAPTER FOUR

Some Common Practices in Apiary Management

It is now assumed that you have by this time got one or two hives which have already been colonised by honey bees. As a beginner, it is advisable to stick to this number for at least one year to enable you to gain a reasonable knowledge of beekeeping before increasing the number of hives.

As you have now been successful in getting bees in your hives, your anticipation is very high as you look forward to reaping a bumper harvest. However, you must remember that success in beekeeping depends on the exercise of your knowledge of colony organization in relation to various factors. It is also controlled or affected by the seasonal and climatic changes, not forgetting the vegetation or the existence of bee forage in the area. A farmer who plants his crop on fertile land with excellent climatic conditions is bound to fail if he leaves everything to chance; thus neglecting other important managerial practices like pest control, clearing the bush, pruning, thinning, etc. Beekeeping is not labour intensive and does not require as much effort as farming does but there are some minor practices which are vital to the survival and well-being of every bee colony. In this chapter your attention is being drawn to the prevention and control of swarming and robbing and to the discussion of a few necessary points such as the provision of food and water for bees, ventilation of the hive and the keeping of records. Diseases, pest control and other problems are discussed in Chapters 6 and 7 respectively.

Controlling Swarming to Your Advantage

In the previous chapter we read about swarming as another

means to increase the number of the existing colonies of bees. You might have been very unhappy about the fact that the precious bees in your hive will one day divide themselves and part of them will leave and settle at an unknown place. You will be more worried about the honey and other valuables that will be taken away from your hive. Swarming divides the population of the colony which causes a considerable reduction of the field working force. As a result, the amount of honey and other valuable products that the colony might produce is considerably reduced. Consequently, the beekeeper would prefer to retain all the bees and make valuable use of them. To do this is to control swarming in a manner that will not interfere with the bees' natural instinctive desire. Such interference could lead to a disaster retarding general output of work or leading to absconding.

Let us now look for the circumstances that lead the bees to swarm. We know that during the peak of the brood-rearing period the best queens are capable of laying up to 3,500 eggs a day. This occurs between August and October, and January to May (these periods may differ in some parts of the region). Thus the brood combs become so populous that such good queens can no longer withstand the congestion of the breed nest. The whole colony is thrown out of balance and workers begin to build queen cells to rear queens for the purpose of swarming. The queen cells are numerous and are built in twos or threes at intervals of two days. This provision is probably made to ensure that an emerged queen will have time to leave before the next one so that peace will prevail in the city. Remember that two queens do not stay in a hive. Indeed queens are reared not only for swarming but for other emergencies such as when a queen dies (queenlessness) or when a queen fails to lay the desired number of eggs. this will prompt the workers to rear a queen to supersede the old queen. Supersedure queen cells, which are few in number and about the same age, are constructed on the surface of a brood comb.

To prevent swarming is to manage the hive in a way that congestion will be minimised or avoided. The idea is to create a commodious area to cope with the ever-increasing brood during the build-up stages. Any managerial activity that will increase the desired 'rooms' for the comfort of her majesty and her

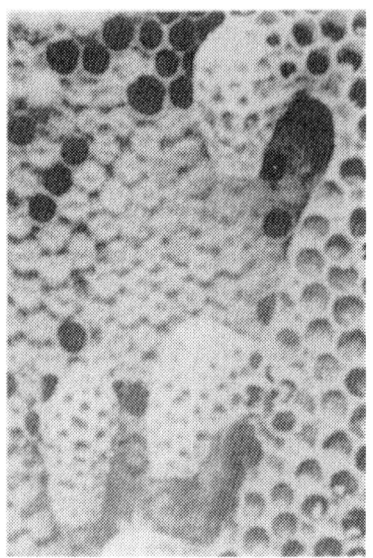

(i) Supersedure queen cells. Notice that these cells are all of the same approximate age. Supersedure cells tend to be large and lavishly supplied with royal jelly. Notice how these cells were built next to the damaged area on the face of the comb. Some authorities tend to the belief that a queen raised under the supersedure impulse cannot be surpassed for quality. As a result of supersedure it occasionally happens that two queens, mother and daughter, inhabit the same hive.

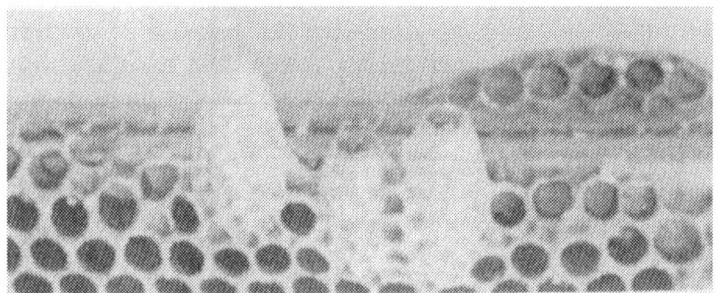

(ii) Queen cells built under the swarming impulse. Notice that in this case there are three cells of varying ages. The cell on the right is already sealed and in the pupal stage. The queen cell on the extreme left is just approaching the sealing stage, while the centre queen cell is still in the mid-larval stage and will be the last cell to be sealed.

PLATE 8: (i) Supersedure queen cells (ii) swarming queen cells (by courtesy of Dadant & Sons Inc).

subjects will undoubtedly prevent or delay swarming. Below are some suggestions which could help.
(i) Honey combs near the broodnests must be removed and replaced by empty combs.
(ii) You may add empty combs (if you can get them) from other hives and place at the sides of the brood combs of the overcrowded hive.
(iii) Provide shade to reduce heat in the hive. Any time you detect that the bees are clustering at the face and sides of the hive is a sign that the inner part of the hive is too warm for them.
(iv) Provide ventilation by placing a stick between the lid and top bars or frames. This will allow heat within the hive to escape and fresh air to enter. They need ventilation if you detect that they are fanning the hive from outside.
(v) Provide water (see Watering bees). Bees use water to cool the hive.
(vi) If you come across some queen cells built at the base or sides of the brood combs then this is an indication that your bees are on their way to swarming. You must intervene to turn this activity to your advantage.

Hiving by Dividing an Established Colony[1]

This book has already given you three methods of securing bees for your hive. They are:
(i) by capturing a swarm in tree branches
(ii) baiting a swarm to enter the hive of the swarm catcher and
(iii) by capturing wild bees by force (i.e. hewing the wood that contains the bees and smoking them out).

Points to Note. Be sure there is abundant bee food available before deciding to divide. Look around to see whether there are bee-forage flowers in bloom. The success of this exercise will depend on availability of food to feed the new colony. Food may be provided by supplying combs containing honey and

---

1. It is believed that there are many species of the tropical bee. Some work better than others so the beekeeper is advised to watch their hives and select those good colonies for this exercise.

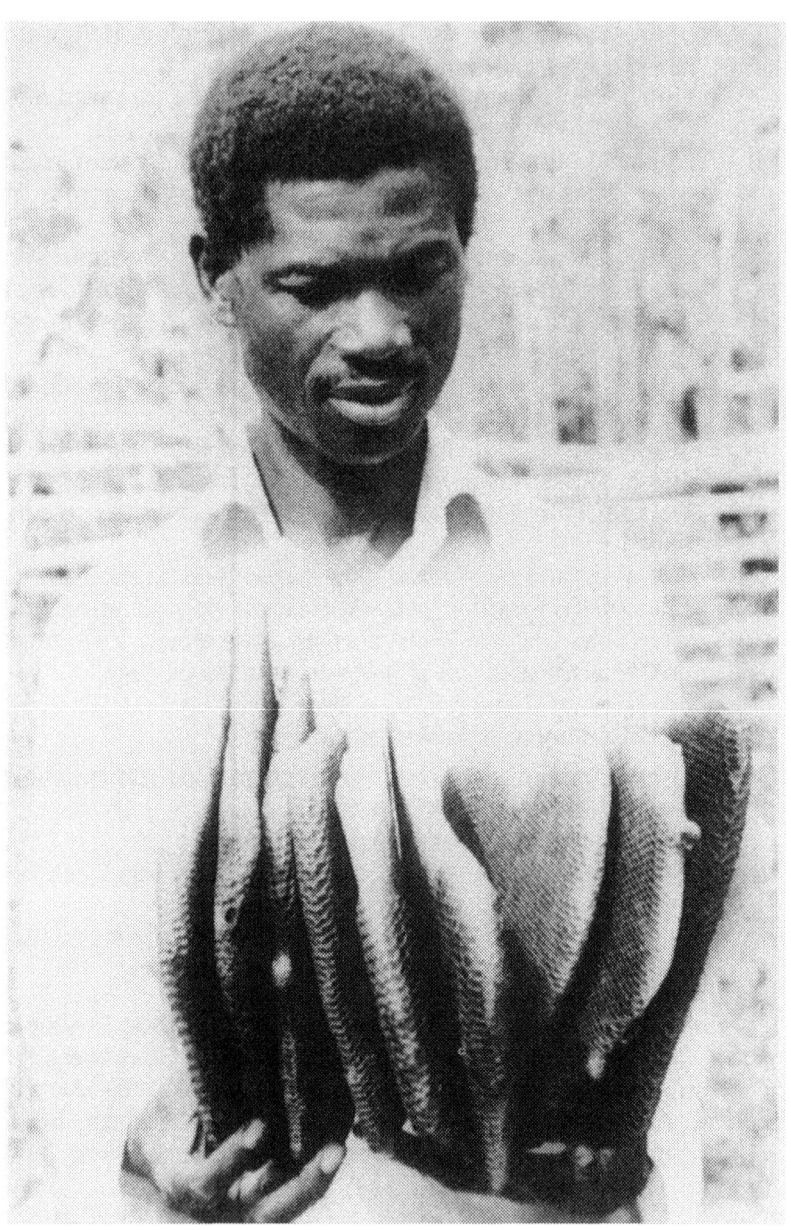

PLATE 9: Combs from absconded hive. Note the swarming queen cells at the side of the combs.

pollen or by placing balls of mashed yam with cassava powder inside the hive. To divide, please follow these instructions:

(i) Remove two or three combs from the old hive which should contain at least five to nine brood combs. Be sure that those you have taken away contain eggs, larvae or capped brood and queen cells if possible. Place them into the new hive. The old queen should remain in the old hive.

(ii) Take out two or more combs; one should contain unripe honey (uncapped honey) and the other should be empty. Put them side by side next to the brood combs.

(iii) Try to scoop more bees into the new nucleus.

(iv) Close up the new hive. Seal off the entrance with any plastic substance so that the bees will remain in the new hive for 24 hours.

(v) Transport the new hive to another spot, preferably at least 3km away. It should remain for two weeks before transporting it back near the old colony. Remove the sealed entrance to allow the bees to move freely. The old field bees will join their colleagues in the old hive if they can find their way back but the young bees will remain to help rear a queen for the new nucleus.

(vi) Visit the new colony regularly for the first month to satisfy yourself that any natural enemies do not disturb them. Do not interfere or harass them by frequently opening and inspecting them in the hive.

(vii) Check whether there is a queen in it (after one month). You will know this by the existence of eggs, larvae and capped brood combs. If you see these do not try to find the queen for you might disturb her. Eggs and brood indcate that a queen is present.

How to Unite Bees or make an Increase

Caution: Under normal circumstances when a honey bee enters another colony's nest, she is regarded as a thief and therefore treated in the same way as human beings treat a thief. The trespassing bee is mauled and carried away by vigilant guards. But sometimes a honey bee may be unable to find her way back to her colony. When she is lost in this way, she will try to obtain permission to join another colony's nest. The guards, upon noticing that she is a strange bee, will try to maul her and the visitor will sometimes have to bribe them by presenting some regurgitated food (i.e. food prepared in the bee's body). Strange bees or bees from another colony are not easily accepted. Therefore should the beekeeper try to unite bees, he or she must bear in mind the following passage.

Bees may be united by:
    (i) adding a swarm to an established colony
    (ii) by adding a queenless colony to a queen-right colony or
    (iii) by joining bees from different colonies to form a nucleus (for establishment of a new colony.).

The beekeeper must remember that a newly-formed colony may be susceptible to chill brood. Please refer to chapter 7 for instructions.

Uniting a Swarm and a Colony

This frequently happens when the beginning apiarist lacks beehives or when he has a weak colony which he wants to strengthen.
    (i) one of the queens of the two, i.e. the established colony or the swarm, must be eliminated at least 24 hours before uniting takes place. Two queens cannot stay peacefully in one hive. They will fight with each other trying to kill the other or the workers will try and kill one of them.
    (ii) Carry the swarm (from their site) to the colony.
    (iii) Using your own judgement, determine which colony is stronger in numbers and add it to the weaker. (Refer to hiving the bees).
    (iv) You may puff some smoke on them to give a new homogeneous smell. Note whether any of the bees died in the process. If there are no casualties then they have accepted each other and will live peacefully.

Uniting a Queenless to a Queen-right Colony:
    (i) Add the queenless colony's bees to the queen-right colony. It is advisable to carry the new colony 3km away. If the hives were adjacent to each other, they will not have to be moved after joining. Do not forget to remove the empty

hive after the exercise.
(ii) Use paper and a quill to unite them as already explained.

Formation of a Nucleus

This necessary activity is performed in order to form a new colony for hiving. It is also a management step when rearing queens.
(i) Shake out bees from two or more different colonies into one hive.
(ii) as the new bees join the nucleus, puff smoke on them to give the nucleus one smell (as above).

When a Colony Swarms

The population may reduce so much that the brood may be left uncovered. If this occurs, more bees must be added to clothe the brood combs left inside the hive. If this is not done, the exposed brood is quickly cooled and will die of chilled brood (refer to Chapter 7). In this case, bees from any source (whether a swarm, queenless colony, etc) but preferably from a strong colony should be collected and united. Always remember that the bees must be moved at least 3km so that they will not find their way to their former colony site.

How to Prevent Robbing

Nature has endowed honey bees with an instinct that guides them to collect sweet juices and to store them for their own use during cool, rainy weather when they remain indoors. Thus the instinct to collect and store is so strong that whenever they locate any sweet juice they regard it as their property. Colonies have no respect for each other when it comes to the possession of honey. They will rob other colonies of their honey at the least opportunity, especially when there is little nectar in the field. Strong colonies with the largest stores are the most aggressive and prey upon the weak ones. It might seem that there is no cause for alarm if your own bees rob each other, but you do have cause to be concerned. If a colony continuously robs another, the victimised colony cannot grow to be a strong colony. Obviously weak colonies will hardly be able to repulse any attack from a swarm of ants or other predators. So robbing should be guarded against at all costs. Some robbing can be carried on so secretly that the beekeeper hardly notices it. The robbers do not enter in large numbers and no confrontation is detected. The robbers sneak through the entrance and cracks, by-passing guards. After taking their fill of honey, they quietly slip out with their treasure. Sometimes the beekeeper can detect that robbing is taking place. A burglary is occurring when a number of bees are flying vagrantly about hunting at all corners and cracks of the hive.[1] But during this time it is not advisable to approach the hive. You do so at your risk as you will never escape without one or two stings unless you are fully protected. Robber bees eventually become skinny and smooth because of their notorious activity. They are always nervous and guilty. Sometimes they cannot alight on the platform boldly and when they are caught by the guards, they pull away. To prevent robbing the following points must be observed:

(i) During brood nest control and harvesting always work speedily and never leave combs exposed. Avoid spilling any honey near the hives as this will attract other passing bees which will create trouble for you and the colony.

(ii) Design your hive so that you can reduce the entrance at will. This is necessary because you will want to protect your weak colony by reducing the entrance and widening it accordingly as their population begins to grow.

(iii) Use repellents such as gasoline and carbolic acid in cracks (if any). This will discourage robbers from approaching the hive.

(iv) During bad weather, you may feed bees in the morning and in the evening by putting the feed only inside of the hive.

Feeding Bees

There are two periods when it is necessary to feed bees. These are the rainy season and the early stages of the colony. There are two rainy seasons in the forest and the transition forest zones of Ghana. The heaviest rains occur between June and July and the later part of September to the first week in November. These

---

1. The writer usually notices this phenomenon between 4pm and 5pm. Nothing is done to stop it. The bees become calm again and settle peacefully when it becomes dark.

two seasons are the most important to mention because the intermittent rains are not so continuous as to stop honey bees from working throughout the day. It is these two periods that a beekeeper in Ghana can describe as a time of famine in his industry. In the peaks of the rainy season it can rain continuously for three or four days with one to three-hour breaks in between. Sometimes when the rain breaks, there is not sunshine and the atmosphere is misty, bleak and cloudy. The bees will never go out in such conditions. Moreover, nectar in the field is so diluted, the sugar content becomes so low that the honey bee does not recognise it. Pollen is also soaked or washed away. It is very difficult, even for the bees who would risk their lives to forage, to collect some little food to send to her majesty's home. So most of the bees remain indoors and feed on honey which was gathered and stored during the honey-flow period. An average colony will thus consume 1.4kg of honey a day and this will go on (unless the rain stops) until all stores are completely exhausted. When such a situation occurs, it is advisable for the beekeeper to feed his bees with sugar syrup,[1] honey, or any other sweet juices like orange, mango etc. The honey that has been produced in the hive is meant for consumption by the bees during the rainy season. But it is just before the rains begin that the beekeeper comes to rob the colony of this very supply.

Leave some Honey for the Bees

You may have now realised how important it is to leave some honey in the hive whenever you harvest your honey crop. Leave at least seven to ten combs containing honey and brood. Do not deplete the hive of every drop of honey. If you do so, then be sure to provide about 1.4kg of sugar syrup for the bees every day until the rains stop. In the previous chapter you have read about how important it is to provide feed for your new weak colony and you have been taught how to prepare your sugar syrup.

---

1. It seems ridiculous to mention sugar syrup to feed bees at a time in the history of Ghana when human beings cannot get it. Sugar is very scarce in the market. But maybe somebody can afford it. If you cannot, leave the bees alone.

Unconscious Method of Feeding Bees

There are many species of tree which are good for the bee pasture. Such trees are listed at Appendix C. If you cannot grow such trees for your apiary it does not matter. But if you happen to live near orchards and such government artificial forests (of bee foraging trees) then take advantage of the opportunity. The Kaduna State of Nigeria is full of such forests of nectariferous trees for bees. It is the duty of the beekeeper to watch when such trees flower so as to know when it is necessary to provide bees with food.

Another way to detect whether or not to provide feed is by weighing the hive. When the hive is heavy it indicates that there is plenty of honey within. When it is light, it indicates emptiness of combs. There is no need to buy a scale.[1] You may use your hands to lift the hive to see whether it is heavy or not. If the hive is hanging on the branch of a tree, you can easily watch how the branch bends with the weight of increasing honey stores.

Watering Bees

We have been told to site our hives near places where the bees will have easy access to a regular water supply. Water is very important for honey bees. They use large quantities of water to dilute their brood food and also use it to cool the hive by evaporation. The need for water to prepare brood food is so necessary that bees have been known to harass villagers in the savanna woodlands and the transition forest zones during the dry season. Water fetched for house-hold purposes can easily be snatched by bees. They will visit watering tanks, stand pipes, pools and sometimes even obscure places such as the urinal, latrine or garbage dump. They enjoy salty water very much. If colonies are not placed near any source of water, it is desirable to provide some. This is especially true during the dry season when water is scarce, otherwise life becomes unbearable in the bee populous areas such as the Brong Ahafo, Northern and Upper Regions of Ghana.

1. Another method which will help you to know what is generally happening in your hives is to keep one glass beehive. It should be covered so that light will not go through but if you want to inspect it, you can look through the glass. This will help you to know the harvest season too. The TCC sells glass demonstration beehives.

Date	Hive No.	Nature of Attack	Remarks
7th June 1981	BG 02	Red ant invasion	—
5th August 1981	AH 15	Moth attack	A red moth, believed to be a type of wax moth, laid eggs in combs.
10th December 1981	AT 01	Bee pirate attack	A wasplike insect was seen molesting foraging bees.

Such records will aid all beekeepers to know more about the local conditions and associated problems. It is also helpful to record the date of every harvest as follows:

Date	Hive No.	Honey	Quantity of Wax	Propolis	No. of Combs taken	Combs Left	Flowering Situation	Temp. at 6 pm
8/3/80	BG 02	9kg	0.7kg	0.3kg	10	8	Old Flowers	28°C
10/3/80	BG 04	14kg	0.72kg	0.3kg	10	8	No Flowers	28°C
15/3/80	BG 03	23kg	0.75kg	0.3kg	10	8	No Flowers	29°C
22/3/80	BG 01	4kg	0.75kg	0.35kg	12	8	New Flowers	27°C

Water can be provided in the manner already described or you may water your colonies as the poultry farmer does.

Keeping Records

Beekeeping in West Africa is now an infant industry. Bee houses are not known and records kept by pasting cards on hives are not applicable. It may be possible to keep records in notebooks. Two main records are necessary and these are:
(a) Colony, and
(b) Operational records

Colony Records

Keep records of each colony. State when and how the hive was colonized. With the help of graph sheets roughly plot the colony's growth. Record everything you notice whenever you visit each hive as shown in specimen record on page 55.

The table above provides very useful information to the Beekeeper. BG stands for Botanic Gardens. This shows that four hives were harvested. Hive No.2 was the first to be harvested on the 8 march, 1980. 9kg of honey was collected from the combs. The flowering situation is also given. On the 15 March of the same year hive No.3 was harvested and from 10 combs 23kg of honey was collected. On the 22 March of the same year though twelve combs were removed from hive No.1 only 4kg of honey was collected. Perhaps some rainfall occurred between 15 and 22. All this information should be included on the table.

The example given shows you how you could make such records. This record is important since you may have nobody in your area to tell you when to harvest and what things to look for and so on. Your own table should also include information on climatic conditions as well as the temperature of a particular hour of the day if you are able to measure it.

Such records will help to pinpoint your good and bad colonies, your most diligent workers and your most productive queens. This information will help you to know which colonies should be selected to build your new nuclei.

Operational Records

This is mostly connected with expenditure and cash flow. This will help you to know whether you are operating at a loss or gain and whether it is worth buying more bee equipment.

Other Managerial Practices

There are a lot of managerial practices which have been left untouched. Some are not applicable in this country and some would only confuse beginners but can be learned later when you are more experienced. As the industry begins to grow it is hoped that this book will be revised regularly. Such important practices will be mentioned in future editions.

CHAPTER FIVE

How to Manipulate Bees and Extract Honey and Beeswax

Why do you have to handle bees?

Two reasons are obvious. They are (i) Harvesting the honey crop and (ii) controlling the brood-nest. Indeed most people keep bees for honey and this chapter explains how to harvest the honey combs to extract raw beeswax for the market.

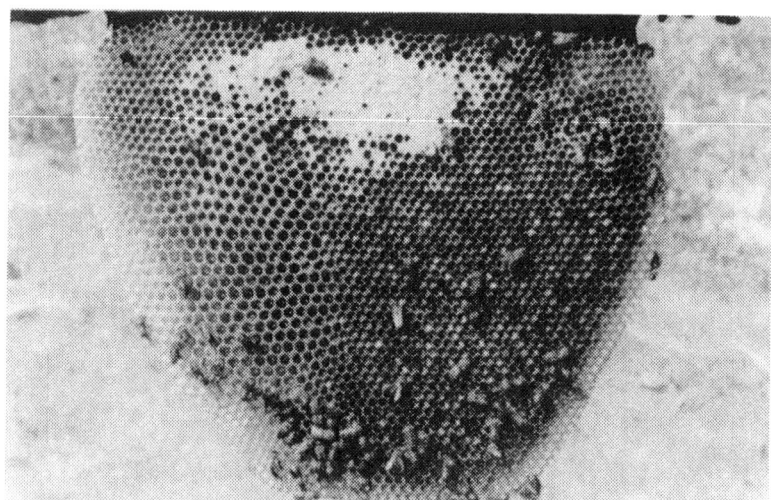

PLATE 10: A honey comb partly capped. Note how the bees seal the ripe honey combs with wax.

Honey Harvesting and What to Look For

How does the beekeeper begin to harvest honey after obtaining a colony of bees in his hive? When does he harvest and how often in one year? These are some of the numerous questions which the beginning beekeepers ask.

Harvesting of the first honey crop depends on the time the hive was colonised by the bees. Like a tree, when a branch of a cocoa tree (for example) is cut and nursed in the soil for the purpose of re-planting, this young tree begins to flower at the same time its parent plant flowers. So it is with a colony of bees. Once its parent colony is capping honey, so will the offspring colony. This is probably because nectar is available; the locality has been experiencing some degree of drought so honey must be collected for use during the dearth season by the bees. For example, the beekeeper is bound to harvest some quantity of honey in the deciduous forest in January/February if his hive was colonised earlier.[1] Some young colonies have already given us surprising harvests. From experience, the first one was recorded at Sunyani in Brong Ahafo where a hive installed right in a backyard garden produced 11kg of honey in 26 days. Another hive at the University of Science and Technology's guest house in Accra produced 15kg of honey nine weeks after colonisation.

These two cases referred to above cannot be desribed as abnormal. It really happens but some colonies are so slow that they need six to twelve months to build and produce adequate honey for harvesting. Three months may be enough for the beekeeper to harvest something, but the quantity of honey will depend on the individual strength of the colony.

It is advised that the first 10 honey combs should not be

1. Two glass bee-hives kept here (at UST Botanic Garden, Kumasi) have shown that bees in our sub-region have two main seasons of the year: each season lasting for six months. Each colony's season depends on the type or strain of bee. One glass hive kept by Mrs Marlene Moshage and colonised in July, swarmed in November and it swarmed again in May of the following year. The writer's glasshive which was colonies in late September, swarmed three months later in early December and swarmed again in mid-June of the following year. Many bee-keepers advise that the new colony should be left untouched for at least one year but this book does not agree with such a suggestion because of the sensitive high swarming instinctive character of the tropical honey bee. Someone remarked that, 'they swarm at random'. After swarming some colonies' populations are reduced so drastically that the beekeeper's attention must be directed to protecting the weakened colony. Thus it is necessary to harvest before swarming to reduce the number of combs to a controllable number so that the remaining bees can guard them. Otherwise the great wax moth will not be prevented from entering and laying its eggs in the hive; upon emergence the larvae will develop and destroy the whole colony.

tampered with. If a young colony has produced more than 10 combs within the period then there is abundant honey and the beekeeper should harvest something.

Harvesting of honey exists between November and June. This does not mean that every bee colony's honey could be harvested. The beekeeper should inspect his colonies and weigh the hives. Some may be ready for harvesting and some may have to be deferred until the next opportune time within the honey period (i.e. November to June). There are minor harvest periods in November, May and June whilst the main season is between January and March. Honey hunters study the silk cotton tree *(kapok)* for the main harvest. They harvest the wild colonies when the flowers wither. In fact the study of the local flowering is very necessary for this exercise. Wait until most of the flowers drop. It is at this time when the bees begin to seal the honey combs.

Other Signs to Note

The best harvesting time occurs during the peak of the dry season when the nights are very warm. Most people feel the need to sleep outside during that time. Do not wait until the windy nights and the first rainfall in the year or you will have missed the prime harvesting opportunity. We have had several experiences of a young colony swarming as early as fifteen days after colonisation but the next swarming takes place six months later. Honey is accumulated before the swarm leaves for their new location.

Develop an interest in the study of the flowering situation of your locality. You may also remember from your theory that the honey harvest comes just before the main swarming season and before a colony swarms it has to make the following preparations:

(i) Building of drone cells and rearing of drones.
(ii) The newly-emerged drones leave the hive to enjoy fresh air in the warm sunshine between 3 and 5pm in the afternoon. Their exercise flights are a common sight for the beekeepers.
(iii) Building of queen cells at the side of the top bar comb and rearing of queens (refer to Plate 9).

Before the swarming fever catches on, brood rearing will cease. This is characterised by foraging bees sending little or no pollen into the hive. The bees become lazy as foraging activity in general seems to have come to an end. Most of the bees, even those at the entrance, are found ventilating the hive. The buzzing continues throughout the days and nights (as those people who have colonies in their ceiling will confirm). They are capping their honey combs at this time.

Not surprisingly, the major occupation of the house bees is honey-packing whilst the majority of foragers send in nectar loads. They are busy preparing honey, their main food for the dearth period which falls during the rainy season. This honey will be consumed by the bees themselves if man and other natural enemies do not steal it.

When you approach the hive, you may smell honey in the air. The guards or the 'security officers' become more aggressive than ever before and they watch the entrance with super vigilance. They will send out patrol officers to go round and attack any potential intruder seen or smelled in the vicinity. You could be a target if you go near them.

The queen starts a slimming course. The nurse bees give her less food to make her dwindle to a smaller size. Sometimes the queen becomes so small that she is unable to be identified as she assumes almost the size of a worker. This weight loss will help her to fly with the convoy to their new destination.

The population of the 'city' is now at its peak. Sometimes in the evening, especially in the later part of October and May, the entrance becomes very congested by a large cluster of worker bees which appears to form a swarm in itself. They continue to watch vigilantly for intruders. Sometimes, this phenomenon will not be found due to the cold *harmattan* wind and the bees will usually sleep in the hive. If large clouds of smoke pollute the sky during the burning season, this will prevent the bees from clustering at the entrance.

How to Harvest Honey and Control the Brood Nest

The processes of honey harvesting and controlling brood nests are the same in the initial stages so there is no need to treat each separately. You may follow the following steps:

(i) Put on your protective clothes i.e. veil, beesuit, a pair of gloves and a good pair of 'Wellington' boots or any similar shoes that can protect your feet.

(ii) Get your smoker, brush or quill and knife or a hive tool. Another thing you will need is a rust-proof container[1] in which you will keep your honey comb.

(iii) Load your smoker with smouldering cow-dung, sawdust, wood shavings or any dry material that can provide you with sustained white smoke to last throughout the operation. Oils, textiles, or rubber materials whose smoke gives an unpleasant pungent smell should not be used.

(iv) Before starting, check that you are properly dressed. Let somebody check and dress you well because if any part of your body is exposed you will be unable to work. Check your veil, especially where it meets your suit at your neck. If any bee enters your veil, even though she may not sting, you will find the whole exercise unpleasant.

(v) Puff some smoke gently round the hive for some few minutes. Later puff smoke at the entrance.

If it is really a harvest period, there will be a large quantity of honey in the combs. Worker bees will rush and gorge themselves with honey and become docile. If you open and remove some top bars or frames, and the bees rush and pounce on you, stop harvesting. It is not yet time. During harvest time a very little smoke makes the bees tame for some time and you can harvest all alone. If you are there for the purpose of control, then it is important to have a co-worker who is also fully protected so that he operates the smoker while you work. A good strong colony will not allow you to work peacefully if it is not harvest time.

(vi) After puffing the smoke, open the lid.

(vii) Use the knife or the hive tool to knock the top bars or frames to determine the occupied and empty sides of the hive. Your clue is 'empty sides make the most noise'.

(viii) Use the knife or hive tool to remove the top bars from the empty side.

(ix) After removing one top bar from the empty side, puff smoke gently to drive the bees to the other far side. This

---

1. Containers used for processing or storing honey must be stainless steel or plastic.

enables you to remove the comb near to you.

If you are there for broodnest control: i.e. to determine if there is a productive queen in the colony (as you read in Chapter 4, Hiving by Dividing an Established Colony), remember that there is no need to find the queen. If you see that there are eggs or brood, then it means the queen is alive. But if you find eggs crammed in the cells then it means your colony is in danger (refer to Chapter 6, Problems).

(x) Remove two or three empty top bars. The first comb you will find is white and therefore new. It may be empty or it may contain some unripened honey. This comb should not be taken. It is for the bees so put it to the empty side of the hive.

(ix) Sometimes a top bar cannot be removed easily. Perhaps it has become fixed by propolis or the attached comb has been fixed to another comb or the hive body. Use a knife to separate it and take it out gently. Remove only *capped* or *partly* capped honey combs; only these contain the ripe honey. Honey combs are always very heavy. Use the feather or brush to sweep all bees on the comb into the hive. Cut the comb, leaving about 2cm length on the top bar for the bees. This will serve as a guide for building the next honey comb. Carry on with your harvest until you come across a brood comb dark in colour and containing pollen too. You may stop here. This is done to ensure that you have left enough honey for the bees to eat during the lean season.

(xii) Use the knife to remove the sticky black propolis from the top bars. Arrange the top bars carefully in the same manner as you found them. If the bees are rushing out between the top bars drive them back with smoke or the quill.

(xiii) Close the hive carefully making sure you have the lid on nicely. The process described here must be executed with swiftness.

Note: The best time to do this is in the evening i.e. after 5.30 pm and in the early morning before 9 am. Any operation or control done outside the time stipulated could lead to a great disaster especially if the hive is not far from a town or village. The

Ghanaian tropical bees are very aggressive.

Experience has shown that some of the bees can attack the nearby villages at the same time the beekeeper is working. On one occasion at Atebubu, a hive hanging in the bush about one hundred metres away was opened for photographing. Alas, all the villagers had to evacuate with even mothers abandoning their babies!

The new beehive with the entrance in the mid-section usually has its honey combs at the sides as shown in the diagram below (Fig.5).

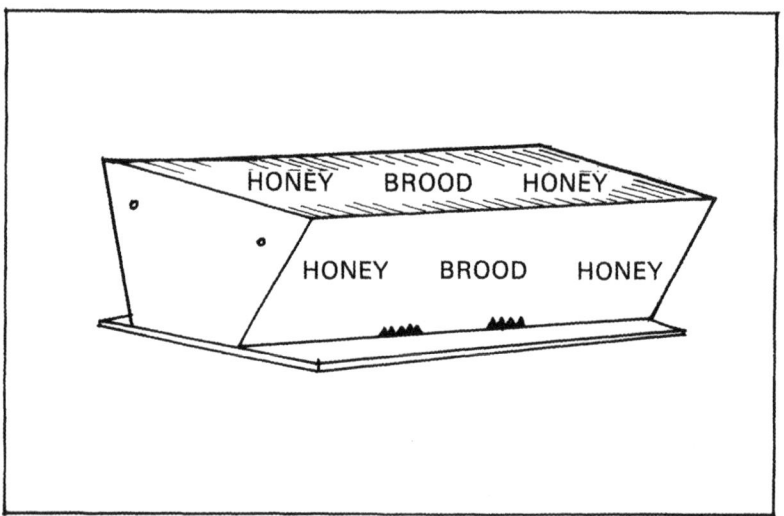

FIGURE 5: Showing the position of honey and brood combs.

At least 10 combs must be left untouched in the centre. These may contain brood and honey (side by side). Leave this honey for the bees to consume.

Bee Sting

Many people are afraid of bees because the tropical African bee is very wild and stings painfully. Bees in the forest areas are not so aggressive as those in the savanna regions; bees near the desert areas are the most aggressive of all. The least disturbance may precipitate a swarm. Bees can kill but the keeper who is afraid of bees could be likened to a taxi driver who would not drive because of fear of an accident; or a farmer who would not go to farm for the fear of a cutlass wound or a snake bite. Such a coward cannot be a beekeeper. However, if you are allergic to bee stings, then do not begin beekeeping. It is interesting to note that some doctors say that bee stings are good and can cure rheumatism. Whether this is true or false, a beekeeper should not be afraid of bee stings. You may avoid it by wearing your protective clothes whenever you visit your hive.

If the bee manages to insert her sting, it should be removed as quickly as possible by scraping it off with a knife or finger-nail. Do not attempt to pick it off. If you do so, you squeeze poison into your flesh. If the result is itching and swelling, do not rub the spot as such rubbing will cause greater pain and swelling. You may treat bee stings by applying cold cloths. In extreme cases victims should be sent to the hospital without delay. Epinephrine could be administered when a doctor's help cannot be obtained.

If you visit a hive unprotected and you are attacked and stung, do not stay there for very long. It is advisable to get away quickly otherwise the odour (pheromone) of the sting will summon more bees to pounce on you. To avoid stings abide by the rules below when visting honey bees.

(i) Wash yourself to make sure you are odour free.
(ii) Do not use any cosmetics as they may contain bees-wax or any smell which could easily cause the bees to attack.
(iii) Approach your hive from behind the entrance or at the sides.
(iv) Do not visit a hive in dark and blue clothes. Always be in white, green or yellow.
(v) Always advance quietly and stealthily. Do not talk or make any noise whatsoever as such noise will agitate the bees.
(vi) Always hold top bars with utmost care so that they will not drop.
(vii) Provide bees with water in drought season.
(viii) Be careful you do not crush any bee. Crushed bees release an alarm scent that alerts the other attackers.

On the Attack

Anytime you have not been attacked it is safe for you to be

there. The first bee that gives you a sting causes trouble which could easily be a disaster for you and the colony. Note the following point well. When you have been stung, the pheromone which the bee has sprayed on the spot has a very powerful smell which summons several attackers to strike. If you stand there quietly without removing the smell, the next attacker will sting right at the same spot. Because of the pain you will not stand still. So as you move, the other bees will find other spots to sting and spray more scent at various places on your skin. This will draw more and more bees to the attack.

Avoid Stings

After the first bee sting you must run away. The bee may chase you but do not be afraid of it because it cannot sting a second time. You may catch and crush it because once it has stung it will die later. Killing it may save you as it will have no chance to go back to the hive and inform others to chase you.

The temper of a bee varies with each individual. Other factors that control her are the weather conditions, the time of the day, the time of the year and the honey flow condition. The operator should always handle bees with care. A queenless colony is very aggressive and restless in the early days. They become very quiet and lazy later.

Extraction of Honey

Better methods of extracting honey are being developed as the traditional methods existing locally are unsuitable and unhygenic. Extraction by squeezing with the hand is the quickest but wasteful and unacceptable. The hand contaminates the honey which causes some honey, such as *neem,* to ferment after storing for a few weeks. The traditional honey tapper's method of burning honey combs to extract both honey and beeswax is totally rejected. The most effective equipment, the centrifugal honey extractor, cannot be used to extract honey from top bar combs. The solar wax melter is now available. But its operation is rather time-consuming in some places (as in Southern Ghana), especially for a large-scale honey-producer. In the absence of a centrifugal honey extractor the solar wax extractor is preferred and can be purchased from local agents involved with beekeeping projects in Ghana.

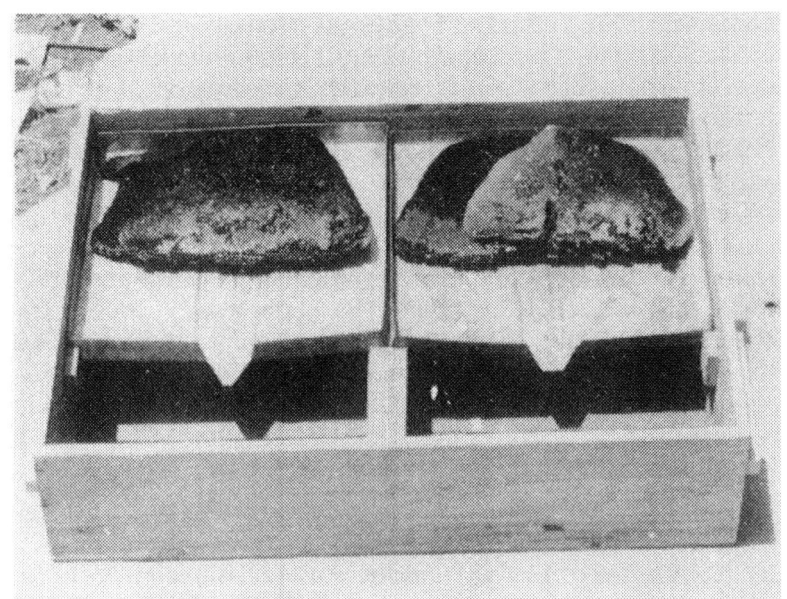

PLATE 11: *Double solar wax extractor. Clear plastic lid removed to show combs in metal pans.*

The Solar Wax Melter

This equipment is made of wood. It has a galvanised metal plate inside and a glass or clear plastic cover. The base is air-tight. It is more advantageous to paint the inside and outside panels (except the galvanised sheet and the glass) black to absorb heat. On a sunny day the wax extractor is capable of generating a temperature of 60°C (or 140°F). This temperature is enough to melt down a bee comb so that both the honey and wax flow into a container which is placed inside the box.

Traditional Method of Extracting Honey and Beeswax

The commercial honey tapper after collecting the ripe honey from tree branches, hollows and crevices, piles all of the combs into a container. Between the container and the honey combs is a metal wire-net. The honey-combs are packed and live embers set on top of the combs. The fire begins to consume the combs. At the same time honey and wax begin to trickle down into the container. This continues until all combs are completely consumed by the fire. The whole stuff is left untouched until the next morning. The owner removes the beeswax which has hardened at the top of the honey. The honey is then poured into bottles. One full beer bottle of pure honey is approximately 1kg and may be sold at any price.

The mistake here is that honey should not be exposed to fire or high temperatures. In the case of the honey tapper, the smoky fire employed is full of ashes, charcoal dust and gravel which contaminates the honey. The honey tastes smoky and bitter. In some cases, brood combs are also burned. The juice of the eggs, larvae, pupae and imagoes goes into the stuff. The watery, contaminated honey is of a very poor grade and begins to ferment within a few days. Such honey cannot be stored for long.

Beeswax Extraction

In the absence of the wax smelter, the hot water bath process now in use in Ghanaian beekeeping villages may be adopted. This method is the quickest way to obtain the wax if the combs are crushed by hand. It is efficient and very effective. Most of the wax from the honey combs is recovered. The pictures from plates 13 to 18 explain the process.

Equipment Needed

    (i) Cooking Pot.
    (ii) A string or twine about two or three metres in length.
    (iii) A stick or a discarded top bar.
    (iv) A ladle.
    (v) A container to hold the beeswax.
    (vi) A container for moulding the wax.

The Process

    (i) Prepare a fire in the same way as you would to cook food or heat water.
    (ii) Put water (depending on the quantity of bee combs) into the cooking pot.
    (iii) Wash crushed bee combs to remove dirt and honey.
    (iv) Put crushed bee combs into the sack-cloth.

PLATE 12: Crushed combs.

   (v) With the twine or string make a good parcel by tightening the string round the sack.
  (vi) By now the water is quite warm and you may put the parcel into it.
 (vii) Upon reaching about 59°C the wax begins to melt down and a waxy scum begins to form on top of the water.
(viii) With the help of the stick (or a discarded top bar) push the parcel into the bottom of the cooking pot. As the parcel is submerged, use the stick to gently mash the parcel. More wax will float to the top of the water.
 (ix) Now use the ladle to skim off the melted wax and pour it into a container. Continue this process until wax no longer rises to the surface.

Note: Do not subject beeswax to high temperatures. Prevent the water from boiling by reducing heat.

PLATE 13

PLATE 14

PLATE 15

PLATE 16

PLATE 17

PLATE 18
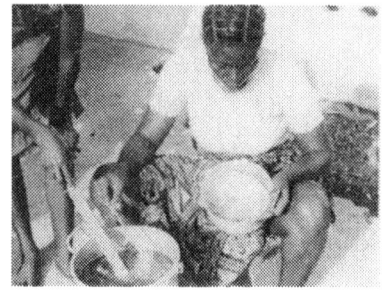

The process of the hot water bath beeswax extraction method now used in Ghanaian villages (photographed by Mrs. M. Moshage).

Moulding the Beeswax

Beeswax collected from the process (as above) must be moulded to obtain a solid mass or cake. This is done as described below. See plates 19-21.

    (i) Find a U-shape container i.e. the mouth is wider than the bottom with very smooth surface inside. A plastic container is preferred. The size of it is determined by the quantity of wax to be shaped.

    (ii) Put a small quantity of water (about one ladle-full) into a cooking pot and put this on the fire.[1]

    (iii) Add all the beeswax. It is important to watch carefully and be sure to remove the pot from the fire immediately after the last lump of wax has melted down.

    (iv) Pour melted beeswax into the mould or plastic container (as at (i) refer to plates 19 and 20).

    (v) Keep the oily beeswax in a cool dry place for cooling to take place.

    (vi) Remove the cakes of beeswax next morning.

Now you have a nice neat cake of beeswax. You will find that all the dirt in the stuff is deposited at the base of the cake. Remove the dirty stuff but keep it for storing. Never throw the dirty wax away. It may be used by craftsmen like shoemakers. The clean beeswax as well as the dirty may be sold to the nearest wax collection agent.

PLATE 19: *The process of moulding beeswax.*

1. Beeswax should not be melted in an empty box. It should not be exposed to fire because it burns very easily and may cause damage. Therefore it is best to melt the combs and mould the wax out of doors.

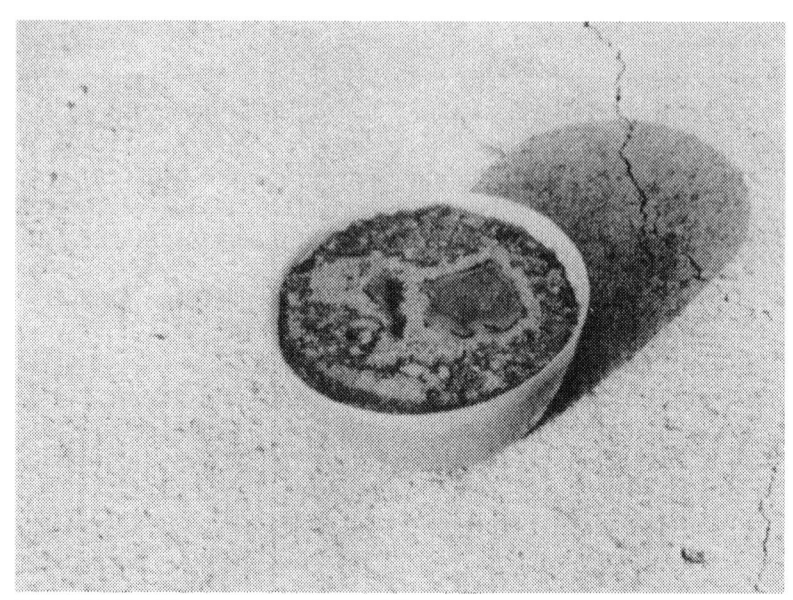

PLATE 20: The process of moulding beeswax.

PLATE 21: Cakes of beeswax.

CHAPTER SIX

Factors Militating Against the Bee Industry in Ghana

In the previous chapters, mention was made of some of the drawbacks of beekeeping in Ghana. These include excessive rainfall in some parts of the country, bush burning and careless handling of wild bees by traditional honey tappers. In this chapter we are going to discuss in detail most of the problems which the bee industry is likely to face and suggest possible ways and means to counteract them. The problems are grouped into four categories.
A. Natural climatic conditions
B. Natural pests
C. Human activities
D. Other miscellaneous problems

A. Natural Climatic Conditions

The amount of rainfall and temperature of an area exert great influence on the life and general work output of the honey bee. Indeed, the honey bee has been called by many authors a 'summer bird' for she is very active and performs most energetically at high temperatures up to 35°C. The activity of the honey bee tends to go down when the temperature drops below 21°C. The honey bee will not move at all at 8°C. Thanks to God, we in the tropics do not know such low temperatures. High temperatures from 36°C and above are equally unfavourable for the honey bee. At very high temperatures, combs begin to melt and most of the bees in a hive will be found frantically fanning themselves and the broodnest. At low temperatures honey bees remain indoors and cluster to generate heat to keep themselves warm. If this continues for a long time, field bees cannot go out to fetch food and other necessities of life. The whole colony will have to depend on honey stored in the cells. The same thing happens during the rainy season. Indeed, honey stored in the cells is meant for consumption during bad weather such as the chilly rainy season.

If confined indoors due to unfavourable weather, one average colony will consume about 1.4kg or 3lbs of honey in a day. If this natural phenomenon continues for a long time the whole store of honey may be completely depleted and the colony will face a famine. The attention of the beekeeper is drawn to this; if he has the means then he should provide sugar syrup for his insects (refer to How To Feed Bees).

Continuous rainfall renders most forest areas unsuitable for commercial honey production.

Such zones include the equatorial forest in the Axim area of the Western Region and parts of the high forest of the deciduous forest zones of the Central, Eastern, Western and Ashanti Regions in Ghana. Ironically, these two forest zones should have been the best areas because of the year-round abundance of flowers, but the excessive rainfall and damp humid atmosphere keep the bees indoors most of the time. Even the periods of dry sunny weather in these areas are inadequate to induce the honey bee to store any appreciable quantity of honey beyond that for 'home consumption'. Added to unfavourable weather conditions and diseases, are the existence of many natural enemies in the rain forests.

B. Natural Enemies and Pests

Ants

The greatest enemies of the honey bee are all types of ants[1] — the black, red, brown, large or small. Before hanging a hive in a tree, first check whether any type of ant is present. If so, avoid it. After hanging or installing a hive be sure to take all the necessary precautions to keep ants out of the hive. This includes coating grease on the strings or cords that support the hive. If the hive has been placed on a wooden table or metal structure,

---

1. Ants have invaded the writer's hives between March and June which coincides with the honey flow period in his area.

be sure to place the stands or legs of the table into a container of fluid oil like gasoil or kerosene.

Weak colonies are unable to repel any serious ant attack so be sure to protect the young weak colonies from this danger. Visit them frequently until they are strong enough to repel any onslaught (for insecticide to repel ants, see section on Poisoning Bees). The pugnacious ant is the most deadly; the beekeeper should vigilantly guard against its attacks.

The Wax Moth (*galleria mellonella* **and** *achroia grisella*)

The wax moth is another dangerous insect. They will attack weak colonies, especially if the entrance of the hive is not well protected or if there are other large openings. The wax moth enters the hive and lays her eggs in the combs. The eggs hatch and the emerged larvae begin to eat the wax, destroying the comb cells. The larvae always spin some webs around themselves; this makes it difficult for the bees to reach and attack them. The moth itself causes no damage. A strong bee colony is able to prevent the moth from entering. Always be sure the beehive has no unguarded holes that will serve as gateways for the moths.

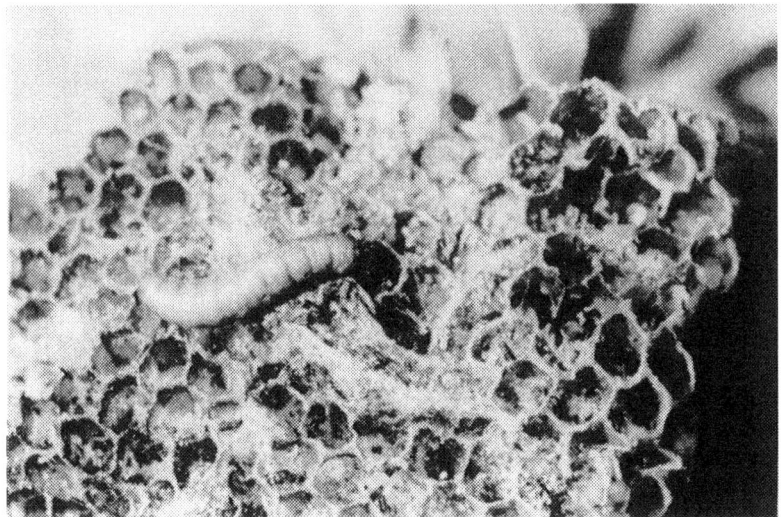

PLATE 22: Wax Moth larva. (Courtesy Mr. Bernhard Clauss).

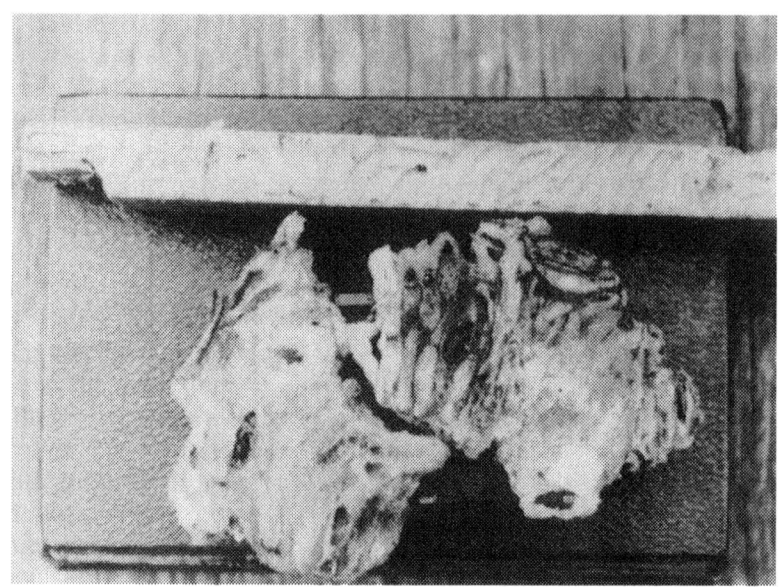

PLATE 23: *Destruction caused to the top bar by wax moth larvae before they turn into pupae. Notice the two clusters of pupae beneath the top bar.*

Acherontia Atropot[1]

This large moth is well known in the forest for entering hives between June and November. Upon reaching the combs, she lays abundant eggs running into hundreds. The eggs hatch and the lavae begin to consume the combs. The destroyed combs will thus collapse onto the base of the hive and decay. Below is a comparison made between the *acherontia atropot*, and the wax moth.

    (i) The wax moth larvae spin webs to conceal themselves and cannot be easily reached by any offended angry bee. The larvae of the *acherontia atropot* do not spin any webs. They produce nothing to shield themselves, but strangely enough the bees seem to give them every chance to destroy their combs.

1. The writer has come across this unusual moth which has just been identified as *acherontia atropot* from the *sphinqidae* family. The new beehives prevent the insect from entering.

(ii) The wax moth activity in the hive does not generate any noticeable stinking material in the hive or around it, but the combs attacked by the *acharontia atropot* larvea collapse and decay; you can smell the abominable scent of the remains up to 10 or even 20m away.

(iii) The number of wax moth larvae found in the hive is far fewer than that of the *atropot* moth larvae which (as I have said) runs into several hundreds.

(iv) The two larvae are similar in size, but the *acherontia atropot* colour is yellowish white.

This unusual moth seems to be more dangerous than the wax moth because its larvae can consume all the colony combs in a shorter time. Alarmed but defenceless, the colony may decide to leave the hive. However, the hive may be recolonized after the beekeeper has destroyed the moths.

The Lizard

The activities of the large lizards found near our homes may greatly concern those who keep bees in their back-yard garden. The lizards stay very close to the hive, sometimes even making the beehive their permanent home. At that convenient spot, they may take their breakfast, snack, lunch and dinner, all of bees! Several suggested precautionary measures are often not effective. Even if the lizards do not reside on the hive-body, they will continue to eat the bees at any opportunity.

They will collect the dead bees which have been thrown overboard by the home bees and the lizards will wait very patiently for this free meal. In many cases, the worker which acts as the 'scavenger' or the 'sanitary officer' will pounce on the old or the sick honey bee. The worker molests it by trying to tear her victim into pieces or by punching holes in her wings. When it is thrown away it will not be able to fly back. The lizard looks on as the unlucky one is being molested. Sometimes the strong bee carries the weaker one out of the hive, puts her on the ground, and then begins to torture her. During the course of the torturing, the lizard will rush quickly and lick the two of them with its sticky tongue.

A serious lizard problem may lead to absconding by the whole colony. Several cases have been recorded. The most feasible of all the protective measures is found below (See Figure 6).

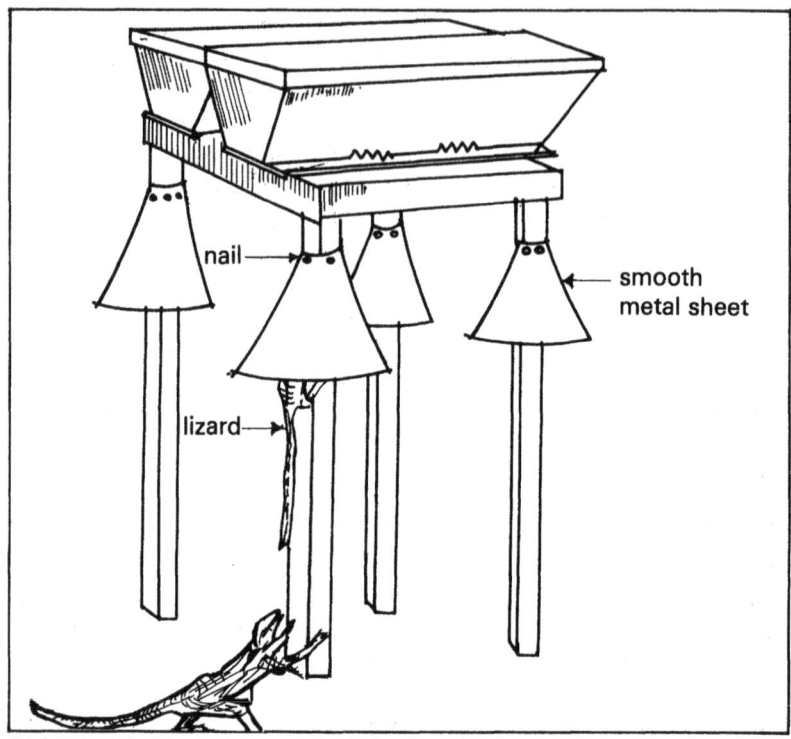

FIGURE 6: Protection against the lizard.

The diagram shows beehives installed on a special platform which has four metal cones nailed on the legs about 70cm above the ground. The lizard cannot climb past these 'guards' to reach the hive.

The Bee Pirate

A wasp-like insect with orange and black skin is sometimes found molesting the field bees entering or leaving the hive. This insect is usually active between October and May. There is nothing the beekeeper can do to stop it.

The Hive Beetle (*aethina tumida*)

These are little black or brown insects with plated shells which the honey bee is unable to crack with her proboscis. They are found in the hive almost every time it is opened, especially if one inspects the cavities of the hive. The bees will try to molest them but their hard shell cannot be pierced by the bees' stingers. The fighting only slows the beetles' determined march to the honey stores. As the bees chase them out, the beetles resist any attempts to remove them. They waste the time of the bees and steal a considerable part of the honey. They will remain in the hive so long as there are bees in it. There is no way to eliminate them from the hive.

The Alpine Swift Bird

This bird is well known for catching and eating bees. These birds come in large numbers during the dry *harmattan* days between December and February. They usually cause a considerable loss of bees to an apiary at this time.

Other Organisms

There are other organisms which follow a swarm and settle with them in the beehive. It is not well known whether they cause any damage to the bee. They are the bee louse and the bee scorpion.

(i) The Bee Louse (*Braula*)

One or two may be found on one worker or drone but more are usually found on the queen bee. This is probably because the *braula* enjoys taking royal jelly, hence it would be the first to partake of the food anytime the queen is served. The bees never attack them. You may de-louse the queen by catching it and placing it between your thumb and middle finger. Place live cigarette ash on the louse and it will quickly fall off.

(ii) The Bee Scorpion (*Pseudoscorpion*)

This, as the name implies, looks very much like a scorpion. It usually clings to the legs of the bee and accompanies them to the nest. The bees try to drive them away but they will never go.

Bee Friends

Other creatures found near the beehive, which do not constitute any danger to the bees, are the praying mantis, the green lizard, wall gechos, some small frogs and lastly, the cockroach. These are usually called the bee friends. They eat some insects which encroach upon the hive, such as the wax moth, house fly, blue bottle fly and the mosquito. However, some people doubt whether the cockroach is really a good friend to the honey bee.

C. Human Activities

In his attempt to improve his living conditions, man has caused — and is still causing — great damage to nature. Bee foliage or forests that nature took millions of years to develop are being destroyed within a few decades. Due to this, the bee population is reducing as their places for shelter are being destroyed through the work of lumbermen, road and building constructors and farmers. Hollows in trees are disappearing as the trees are cut, thus forcing most new bee colonies to hang from tree branches which exposes them to all their natural enemies.

The Honey Hunter

The activities of the traditional honey tappers (our present main supplier of honey) raises a great problem. They make torches to burn the bee colony, rendering them weak in order to harvest the honey. In the process, thousands of the bees are lost and the overwhelmed colony whose 'house' is destroyed has no choice but to abscond in search of a new home. This barbaric way of honey collecting is still going on without any interference. The Government is the only body which can, through the Beekeepers Association, hit this practice by educating the honey tapper.

Bush Burning

In the transition forest zones and the savanna grasslands, bush burning is rampant during the dry season which extends from November to May. Some obvious reasons for the bush burning are: (i) to clear the land for farming (ii) to clear the bush to make hunting easy. The honey bees suffer greatly from such fires as they are burnt to death in large numbers. Imagine that you are keeping hives which you bought. What would be the fate of the hives in a wild fire that can consume an area of 100 square miles?

SOLUTION

In a square mile, group about 30 to 50 of your hives in one area, probably at the bank of a river or a stream. Make a fire belt around the hives during the dry season. Visit the area as frequently as possible checking for fallen wood or leaves which would lead to a fire on your apiary site.

Bee Burning

In the dry season, the honey bee makes life difficult in the densely bee-populated areas. When the land is dry, the streams disappear. Man has to travel miles to fetch a head-load of water for domestic purposes. At the same time the temperature is high. The bees become very aggressive. The hive becomes too warm and needs to be cooled. Fanning alone (by bees) cannot bring the needed relief so a large number of the foraging bees have to fetch water. In large numbers, they will lay claim to that bucket-full of water which the villager has collected. On returning from the farm late in the evening, the exhausted farmer finds that his only bucket-full of water has been drained away by the bees. Added to this, the bees harass women and children pounding grains. In the extreme cases they try to suck human sweat and this results in a scuffle.

The Palm Wine Tapper

Another source of sweet and refreshing liquid for the bee is the palm wine tapper's pot. From observation the honey bee is one of the creatures that awakens very early. It seems that the tropical honey bee starts work earlier than 5 a.m. This is well known by the palm wine tapper who usually makes sure to reach his wine very early. By the time the wine tapper removes the first pot of wine from the 'dead' palm tree, many bees have already been there, filled their stomachs with wine and become tipsy. The bees are unable to take their booty home. The wine tapper, sometimes becomes very furious at the sight of countless bees lying in a deep coma in his wine. He collects all the bees and throws them away. In extreme cases, the whole pot of wine is consumed by bees. The wine-tapper assumes that the motionless tipsy bees are dead and therefore does no further harm to them. However, all the bees left unmolested will recover and fly back to their nests later. This theft, added to all the harassments referred to above, may be the cause underlying bee burning in the transition forest country.

SOLUTION

The surest way to prevent bee-burning is to provide regular water supply for bees and human consumption in the dry season. It is strongly suggested that anyone planning a beekeeping project in the savanna area should not forget to provide adequate water for bees and for the people. This will prevent any bee attack from causing any future loss of life.

Poisoning Bees

The honey bee is popularly known as the chief pollinator of crops. This function of the honey bee is performed unconsciously during her search for nectar and pollen sources. It flies from one plant to the other and from one flower onto another. Sometimes the honey bee unknowingly lands on a poisonous plant or contacts a poisonous insecticide which the farmer has sprayed to protect his crops from dangerous insect pests. Pollen collectors may carry the poisonous pollen into the hive and store it there for future use as bee bread. So long as the poisonous pollen remains in the cells, a dangerous threat is posed. It may kill both adults and brood, either upon contact, or by ingestion of the tainted food by the brood (the adult bees feed on honey and not pollen).

Beekeepers are advised to keep their hives away from pyrethrum plantations. The use of DDT to spray cocoa and other crops cannot be controlled but the beekeeper must be aware that it poses a great danger to the field force of his hives. Insecticide may be needed to repel hive invaders like ants.[1] The only known insecticide that can be used is opigal 50 DDT (dichoro-diphenyl-trichloroethane). But it is lethal to bees and must never be applied near a hive. Another insecticide, DDD

1. The writer always uses opigal 50 to fight ants which invade his hives. This insecticide is effective as a contact killer for most insects, especially those that suck blood. No honey bee has ever suffered from application of this chemical. A substitute for opigal 50 is Asuntol 150. Asuntol is highly dangerous when applied in any form near the hive. It was used when there was no opigal 50 and the entire population, the best colony for the writer at that time, was completely wiped out.

(dichloro-diphenyl-dichloroethane), is said to be less toxic to bees. Large quantities have to be applied before a bee can be killed by contact. DDD is, therefore, a safer insecticide that can be used to repel ants.

D. Other Miscellaneous Problems

Queenlessness or Unfertilized Queen

You have been warned to visit your hive frequently when it is newly colonized by bees. This is necessary to check (i) whether some hive invaders are threatening them and (ii) whether brood is present and healthy. Brood nest and population growth must be checked. If the beekeeper discovers that the colony is not growing then it is possible that the queen is unfertilized or dead.

An unfertilized queen may be the result of prolonged bad weather which prevented the queen from mating. If after about four to six weeks the colony has not grown and the eggs are crammed in the worker cells, then you know that the colony is threatened. The eggs found were laid by the workers and will hatch drones. You must act quickly by inserting a brood comb with newly laid female eggs, or preferably, a comb containing unemerged queens. Be sure to remove the barren queen as she may fight the new queen. The workers will rear a queen to supersede the old queen.

Absconding Colony

One of the least desirable characteristics of the tropical honey bees is their tendency to abscond. In our decidiuous forest, the months noted for this activity are November and May.

A strong, busy colony may suddenly stop carting pollen. Foraging activity (in general) diminishes and the bees begin to drain the honey stored in the hive. The queen stops laying and the colony waits for some time until most of their brood emerges. They then finish the honey and chew through some of the combs. This is probably an attempt to distort or destroy the combs so that no honey bee colony will use the nest again. Such combs may be removed and melted for beeswax. They should not be left for the wax moth larvae to destroy them and to destroy the hive body in addition.[1]

Sometimes the cause of the absconding may be traced to unnecessary harassment or a type of disease. Dirty combs (ie brownish in colour with some unusual dirt) may suggest some virus troubles which might have led the colony to leave. Such bee combs should never be inserted into any other beehive but must be melted for beewax. Combs left in beehives must be fumigated or else they will be destroyed by the wax moth larvae. If fumigants are not available, then do not try to store the combs. Always clean and re-bait the hive after absconding.

PREVENTIONS

Carefully watch your colonies during the months mentioned, but do not disturb the bees. Any disturbance in the preceeding months, i.e. October or April, may cause you a great disappointment. Allow the bees to work. Once they are regularly carting pollen into the hive, they are alright and must not be disturbed. Do not smoke them frequently.

Always inspect your apiary after a rainstorm to make sure that no hive has tumbled over or fallen down. Remember to work in the cold mornings and evenings. Avoid the warm hours of the day when the bees are busiest.

The Spider's Web

Any time you come across a spider's web around or near your hive, destroy it. This can arrest the foraging bees and kill them. Spiders usually make their traps near beehives to catch the bees.

---

1. These hives usually take more time to colonise. It is believed that the bees mark them and this marking discourages swarms from using them.

CHAPTER SEVEN

Some Diseases of the Honey Bee

Like other living creatures, the honey bee has several diseases. Some are known to man and some are still unknown. Investigations are continuing in other parts of the globe to identify them. In our tropical region, where beekeeping is an infant industry, not very much is known but it is believed that most diseases identified in the temperate regions and elsewhere also prevail here. Therefore, a judicious study must be undertaken here in the tropics to identify and combat such diseases if we want to get maximum benefit from our efforts.

From the previous chapters, we know that the honey bee starts life as an egg, then passes through larval and pupal stages before it emerges from its comb-cell as a young honeybee or imago. During the brood stage, the honey bee may suffer from several diseases. We will first concern ourselves with some of these brood diseases. Most of these diseases are believed to be most common in the forest where humidity is high.

The Brood Diseases

The beginner cannot identify a brood diseased comb unless he previously knew what a healthy comb looked like; therefore the beekeeper must begin to study bee combs very carefully.

A healthy bee comb containing brood is usually clean. Good queens lay their eggs in clean cells, one egg per comb cell. The healthy larva coils like a 'comma' in the cell. After the fifth or sixth day, it is capped or sealed with wax. The young larva is a glistering white and is fleshy in appearance. It does not wander or move from place to place in the cell. It does not look black, brown or assume any colour except white. The regularity of the brood in the cells must also be noted. Good brood comb cells are usually compactly filled by the fifth and sixth days before sealing takes place. Therefore an irregular brood-comb must be watched for identification of the brood disease. Care must be taken, however, to judge whether the irregularity is the result of the emerging brood. Pupae must remain capped; the seal should not be punctured or sunken. Any of these irregularities should suggest that something has gone wrong because of a disorder or disease.

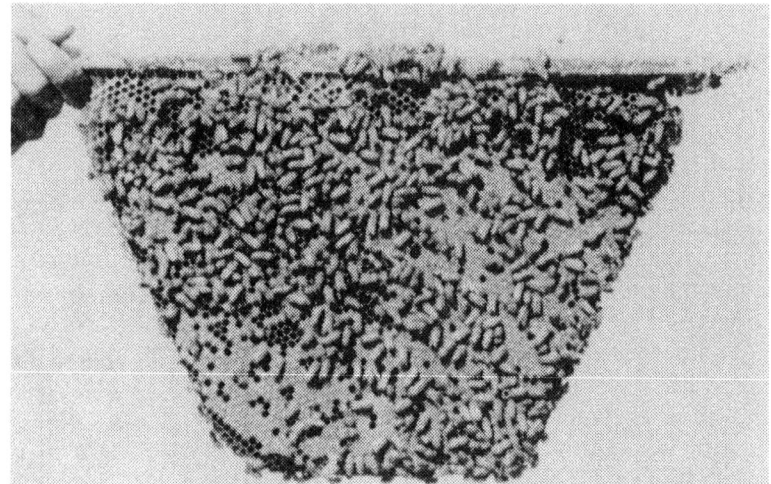

PLATE 24: Emerging bees. Notice the compactness of the capped brood on the right side of the brood comb.

In its developmental process from egg to pupa the honey bee is susceptible to several diseases ranging from a serious one which destroys an entire colony to a mild attack which causes only a small loss of the total population. It is therefore necessary for the beekeeper to study these brood diseases so that he knows what to do when such troubles occur.

The diseases to watch for are:
— the American foul brood;
— the European foul brood;
— sacbrood.

Other abnormal conditions of the brood which are caused by infection are:

— the chilled brood;
— the starved brood;
— the bald brood.

The American foul brood is the most serious of all the brood diseases, followed by the European foul brood. They are so called not because they are peculiar to each of the two continents but the two diseases were studied separately by American and British researchers. The disease identified by the Americans was named after that continent and so the other disease was named after the European.

It is believed that the American foul brood is not present in our region but the European foul brood does exist in our forest; however, it would be unwise not to study them both.

The American Foul Brood

This disease causes heavy losses to the colony's population. It can wipe out not only a single colony but all the colonies in an apiary. It is not seasonal and may occur at any time.

It is caused by bacillus larvae. The bacillus form strong resistant spores. The organisms attack the larva which dies after it has been capped during the pre-pupal stage. The pre-pupa dies, becomes brown and finally dries up into a hard scale which is difficult to remove from the comb-cell.

The normal convex capped cell becomes sunken and sometimes perforated. The decomposed brood has an unpleasant smell. When a small stick is thrust into the cell of the decomposed pupa, it draws out a ropy thread several centimetres in length. The perforation of the capped cells may be the result of the shrinking of the dead pupa or the attempt by the workers to uncap it for removal of the decomposing remains.

ACTION

Call for the officer in charge of beekeeping or the Director, Technology Consultancy Centre, University of Science and Technology, Kumasi, Ghana. If such contact is not possible, then the whole attacked colony must be burned with the top bars, and the debris buried deep in the soil. Drugs like sulfathiozole and oxyteracycline are said to be used, both as a preventive and as a remedy; these are not always readily available. The American foul brood has not been identified in our region but beekeepers should be aware of its characteristics in case it should occur. If the characteristics mentioned above are ever detected, then the beekeeper should make every effort to report the outbreak to the TCC.

The European Foul Brood

This disease is caused by an infective bacteria. The young larva affected dies on the fourth day. In the early stages of the disease the most easily detected organism is the *streptococcus pluton*. This is believed to be the primary causative agent but the larva's death is also accelerated by the presence of *bacterium eurydice*.

SYMPTOMS

The larva that usually coils motionless in its comb-cell begins to wander out of pain. It becomes very restless before it dies. The young larva is infected by taking in food containing bacteria. The bacteria multiply in its gut and feeds on the food taken in by the larva. The larva then dies as a result of starvation on either the fourth or the fifth day.[1]

The position of the dead larva remains unnatural — lying across the mouth of its cell and twisted spirally around the walls or stretched out lengthwise from base to the mouth of the cell. The dead larva is porridge-like in appearance, as if it has melted. Its plump, fleshy appearance is completely lost as it turns yellowish-brown and eventually dries up into brown scales. Sometimes, some of the sick larvae are sealed in the cells and may be seen lying in sunken capped cells. On a normal comb the different stages of brood appear in concentric bands. The queen begins laying eggs from the centre of the comb and works in circles until she reaches the limit of the brood space. When the disease occurs, the regular pattern of egg laying is broken and different age-groups are scattered over the comb. The smell of the decomposed larvae varies according to the species of secondary bacteria which invaded the dead larvae.

1. The writer has discovered the presence of this disease in the Kumasi area. Accra, which is in the coastal scrubland, seems to be free of this disease. Colonies found in Accra are very healthy and produce good brood combs.

CURE

Since the death of the larva is caused by hunger it seems the abundant supply of food to the larva by the nurse bees may solve the problem. Once the larva enters its pupal stage, it will need no more food. In our forest, the disease has been found between June and August. The disease begins to diminish when the colony's population increases considerably in the latter part of August. In October, the affected colony looks normal again but honey yield will be lower than expected. The queen of the affected colony must be replaced. This can be done by transplanting a comb with a capped queen cell from a healthy colony. The population of the affected colony should be increased by uniting it with a swarm or a queenless colony.

Stone brood

This brood disease is caused by a virus called *morator aeratulae*. As with American foul brood, this organism kills the brood in the pre-pupal stage. The brood, in its pre-pupal stage, lies stretched out lengthwise in the sealed cell. After the insect's death, the cell is partly or fully opened and the dead removed from the hive.

Before decay begins, the diseased pre-pupa's sac-like outer skin becomes loose and is filled with a clear liquid. This is clearly seen at its tail end. The head and thorax becomes darker than the other parts of the corpse with the head usually showing black patches. No particular odour is present.

The virus is spread in the nest by the house bees which usually have to evacuate the affected dead brood. The virus does not live for long and the disease may disappear during the honey flow period. Serious out-breaks are not common and usually no controlling action is necessary. If control is considered necessary, then the colony must be requeened.

Nosema-like protozoan[1]

This disease is thought to exist in our forest. It is usually found between April and June. It attacks the brood and kills the

---

1. The writer has several times watched bees disposing of dead pupae affected by this disease in Kumasi.

pupae. The spores left on the dead pupae are similar to those of *Nosema-apis* which attacks the honey bee. The home bees readily remove the dead pupae and throw them away. Sometimes some pupae may be seen lying outside the entrance on the landing board. Colonies may abscond if heavily affected. Those that remain in place may resume normal activity at a later time.

The Chilled brood

This is sometimes called the over-heated, chilled or starved brood. It is not a disease but an abnormality caused either by cold or over-heating.
  (i) When the colony's population declines, fewer home bees will be available for protecting the brood combs. Some brood will be exposed to the air. Remember that a temperature of 35° centigrade is required in the brood chamber. If this is reduced, then the brood suffers. If it continues for hours, the result is that eggs, larvae or pupae can be destroyed.
  (ii) Over-heating also causes a similar negative effect. When the inner part of the hive is over-heated, home or nurse bees will come outside to fan air for themselves thus leaving the brood uncovered. Chillbrood may also result after the colony's population has diminished due to poisoning. The few that are left after poisoning may not be enough to clothe all the brood in the brood-chamber.

The obvious way to combat this is to add more bees to cover the brood combs.

Bald Brood

In Chapter 6 we discussed the problems caused by the wax moth larva. It eats the wax and makes tunnels through the combs. During the process of tunnel-making, sealed brood may be encountered and uncovered. This untimely exposure may cause the pupae to die. Sometimes the web made within the tunnel may fix the pupa in its comb-cell thus making it impossible for it to emerge. The pupae may also dry out to a certain extent and emerge with shrivelled wings or malformed legs.

Genetic Faults

Occasionally, a queen's eggs may not hatch or its pupae will die or never emerge. This may be caused by inbreeding and the only way to overcome this problem is to insert a new queen from another good colony. To avoid problems in requeening it is best to insert a capped queen cell and not an emerged queen.

Diseases of the Adult Bee

Nosema

This is caused by the protozan *Nosema apis*. The bee takes in food which has been contaminated with *nosema* spores which germinate in the mid-gut of the adult bee. A long filament which is like a thread is sent out which penetrates the cells lining the gut. The organism multiplies and most of the cells are infected. The affected bee cannot utilize the body's protein reserves and consequently very little royal jelly or brood food can be produced. Therefore only 15% of the potential number of brood can be reared at a time.

The disease causes the young bee to forage earlier and its life-span is greatly reduced. The water in the body of the affected bee becomes greater than normal and the bee becomes lethargic and may begin to soil the hive. It later becomes a crawler and subsequently collapses.

The affected queen stops laying or the few eggs that she does produce will not hatch. Its life-span is noticeably shortened by the disease. Since the beekeeper needs a microscope to detect this disease, it makes it impossible for the average man to diagnose it. The only visible sign is that the colony becomes weaker and weaker as bees fail to build up when conditions are favourable. The beekeeper may also notice bees crawling about in the hive with swollen abdomens.

CONTROL

The number of bacteria in the brood nest must be reduced. Unfortunately this job cannot be easily accomplished with the top bar hive. New combs must replace those of the old brood nest and the colony should be re-queened. Another remedy is to feed the affected colony with the drug fumagillin at the rate of

100mg active ingredient in a 4 litre solution of one volume sugar to one volume water per colony. If the prescription given here cannot be obtained, the only option left is to burn the colony so that the disease will not spread into other colonies.

This disease is here and all beekeepers must watch for its occurrence in their apiaries.

Dysentery

Adult bees are seen soiling the hive near the entrance or in any other part of the hive. The faeces are yellow. This condition is caused by excess water in the alimentary canal. It is caused by confining bees in the hive for a long time due to bad weather. When this happens, they cannot go out to defaecate.

When bees consume unripe honey they may develop dysentery. Dysentery may kill most of the affected bees. When the colony's number is reduced so dramatically, increase its population by adding a swarm or nucleus colony at a time when the weather is favourable to the bees. The beekeeper must wait to restock until after the disease has become inactive, otherwise the new bees will develop the same condition.

Paralysis

Paralysis is caused by a virus. Two kinds of virus are known to cause paralysis in bees: one is acute and the other is chronic bee paralysis virus.

The bee cannot fly and trembles. It dies after a day or two. The affected bee is usually molested by the other bees. They will sometimes carry it out and throw it several metres away from the hive. Control is not necessary.

There are several other diseases which cannot be dealt with here because their control is difficult and seems complicated to the new beekeeper. As we do not claim to be an authority in this area, we will be very pleased to receive your letters and reports concerning new developments and discoveries in your locality. This will help us to make further investigations, compilations and conclusions so that most of the serious problems will be collectively solved by those involved. It is important that we should not wait for foreigners to work out our problems for us.

Conclusions

The author would like to remind readers that he has never been happy with those who waste their precious time by attending a beekeeping course at the University of Science and Technology, Kumasi, but do nothing with their knowledge. The most serious offenders of all are those who purchase beehives and do not install them simply because they are waiting to acquire a piece of land. The beekeeper does not necessarily need to be a land owner. Such people wait until honey bees colonise their roof before they realise the need to put up their beehives.

The potential of beekeeping in our land is great and untapped. Honey-making is money-making, so make money with honey. As this book comes to an end, readers must be reminded once more that the Golden Insect is neither a history nor a fiction. It must be acquired, read and backed by action.

I hope that this message will be respected and given its due attention. I wish all beekeepers, be they old hands or new beginners, rich honey-flow seasons and good luck.

Appendix A The Uses of Honey

As human food

As sugar substitute
In cooking and baking
In child and infant feeding
For athletics and strenuous activities
For diabetes

As ingredient in drugs

For hayfever (pollen + honey in ratio 1:11 parts)
In cough syrup
As sweetening agent of drugs especially for children

For animal feeding

To feed the dairy cows in order to produce more milk
To feed racehorses and donkeys
In poultry mash
To feed fish in fish farms

In veterinary medicine

In the treatment of *acetonemia* (a disease of cows)

Other miscellaneous uses

For beverages of good alcoholic content
As a facial cleanser
As hand lotion (an ingredient)
Use in mice and rat repellent drugs
Honey has successfully been used as a shock absorber in model A Ford Cars
In curing pipe bowls

As an ingredient of cigarette and chewing tobaccos to improve flavour and texture
Used to keep chewing gum from drying out
To cure hams
To bait honey bees to colonise a hive. This is used to coat on the top bars only i.e. between the lid and top bars so that the bees cannot reach it and consume it.

Appendix B The Uses of Beeswax

Adhesive compositions, ingredient of

Adhesive for wigs and masks
Adhesive for setting bristles in brushes
Adhesive for sealing closet bowls
Adhering metal to glass and glass to glass
Electrical and chemical cement

Candles

Raw material in making candles
Ingredient of
a. Liturgical candles for use in the church
b. Nonrubrical candles containing from 15% to 35% beeswax
c. Sanctuary lights — contain up to 51% beeswax
d. Candle decorations

Comb foundation

Brood foundation
Thin surplus foundation
a. Bulk comb honey
b. Comb honey

Cosmestics — usually in the bleached form as an ingredient of

Actor's grease paint
Camouflage creams and ointments
Cleansing creams
Cold creams
Eyebrow pencils
Lip pomades
Lipsticks

Massage creams
Moustache wax
Paste rouge
Theatrical cream

Crayons, ingredient of

Drawing pastels
Grease pencils
Lithographic crayons
Wax crayons

Dental purposes

Evan's cement
Horsley's wax
Impression wax
Pink base plate wax
Sticky wax
Temporary tooth filler
Toothache gum

Electrical purposes, ingredient of

Filler for transformers and terminal boxes
Insulating compositions for various purposes
Insulating agent in making cables and electrical apparatus

Food, ingredient of

Artificial foundation of comb honey
Chewing gum
Compositions for coating candies
Compositions for decorating fancy foods

Ink ingredient

Lithographic inks
Offset and non-offset compounds
Printing inks
Stamping inks
Transfer inks
Writing inks

Leather, ingredient of

Dressing compositions
Finishing preparations
Various polishes

Metallurgical purposes, ingredient of
Coatings on ammunition and shells
Compositions for preventing rust and corrosions of acids, alkalis, and other chemicals
Electroplating compositions
In shell loading
"Lost wax" process of Benvenuto Cellini for casting statuary in metal
"Lost wax" process applied to industrial castings
Protective agent in making acid etchings
Soap solutions for drawing
Waterproof coating to prevent salt water corrosion

Miscellaneous uses
Basketball molding
Cartridge wax and grease
Composition for minimizing shrinkage of wood
Composition for polishing and cleaning wood or rubber
Foundry pattern making
a. Modelling wax
b. Thin wax sheets
c. Wax fillets
Gilders' wax
Grafting wax
Imitation fruits and flowers
Ironing wax
Modelling wax
Polishing telescopic lenses
Poultry — for removing feathers
Production of acid bottles
Sealing wax
Ski wax
Snow shoe wax
Substitute for paraffin in waxing paper
Waterproofing agent or ingredient of
a. Asbestos composition
b. Coatings for bricks and stone
c. Hat straw
d. Ingredient of artificial stone
e. Porous building material
f. Straw board
Waxing archers' bow strings
Waxing threads in sewing
Wax putty for stopping leaks in casks
Wax soaps

Oils and fats, admixture for special lubricating purposes

Axle grease
Special lubricants
Various gun lubricants

Paints and varnishes, ingredient of

Anti-fouling paints
Lacquers for flexible materials
Paint and varnish removers
Preparations containing dry colours
Various paint mixtures
Varnishes
Wood fillers

Paper, ingredient of

Coating composition for washable wallpaper
Compositions for manufacturing carbon paper
Emulsified sizing preparations
Preparations for waxing paper
Sizing for high-gloss paper
Waterproofing compositions

Pharmaceutical preparations, ingredient of

Almond balls
Brushless shaving cream
Camphor ice
Cerates of various types
Depilatories
Hair restorers
Oxyeroceum plaster
Pomades and "hair straighteners"
Pomatum for chapped lips

Polishing and cleaning preparations, ingredient of

Automobile polishes
Compositions for colouring and polishing wood
Floor oils and waxes
Liquid floor wax
Polishes for automobile tyres
Powdered wax for dance floors
Preparations for cleaning and polishing furniture
Shoe creams, pastes and polishes
Various compositions for cleaning and polishing floors

Printing, ingredient of

Acid proof coatings for plates in electrotyping
Matrices in galvanoplastic work
Process material in
a. Electrotypers' wax
b. Lithography
c. Photoengraving
d. Process engraving

Textiles, ingredient of

Assisting agent in stretching cellulose acetate filaments
Compositions used for finishing
Compositions used for sizing
Compositions used in the manufacture of waxed cloth
Impregnating and coating agents
Various emulsified dressings
Various water proofing compositions
Waterproofing agent in treating yarns and fabrics
Waterproofing canvas
Waterproofing cellulose fibres
Waterproofing threads for shoe and harness making

From: *The Hive and the Honey Bee* pp531-534. Revised Edition Second Printing 1954. By courtesy of Dadant & Sons Inc.

Appendix C Physical Properties of Beeswax

In connection with industrial uses of beeswax the following data may be used from time to time.

Density .... 0.95 gm/cm^3
Melting Point .... 61-64°C
Freezing Point .... 60-63°C
Coefficient of thermal conductivity .... 2.5×10^{-3} $\dfrac{\text{Jcm}}{\text{cm}^2\text{s}°\text{C}}$

Appendix D Honey Bee Forage Trees in Ghana Forests

Botanical	English	Twi or other Vernacular
*Acacia Albida		Gawo (Hausa)
Acacia dudgeoni		Gosei (Kusasi)
Acacia Gourmaensis		Sareso Akoobowere
Acacia nilotica		Gorsia
Acacia polyacantha	Gorpila	
Adansonia digitata		Odade (e), Ototowaa
*Albizia Lebbeck		
Anacardium occidenalis		Atea
Annona senegalensis		Saa-borofere, Aboboma
Anogeissus leiocarpus		Kane
Azadirachta indica	Neem	Amanyedua (Twi) Kintso (Ga)
Eutyrospermum parkii		Kra-nku
*Cassia siamea	Cassia	
*Casuarina equisetifolia	Whistling pine	
Ceiba pentandra	Silk Cotton	Onyina, Onyaa
Citrus aurantifolia	Coconut	Ankaa, Akutu
Cocos nocifera		Kokosi, kube
Combretum paniculatum		Ohwirem
Dichrostachys gloromerata		Akyekyere-besi
Diospyros mespiliformis		Okisibiri
*Eucalyptus alba	Eucalyptus	
*Eucalyptus cadambae	Eucalyptus	
*Eycalyptus Citriodra		
*Eucalyptus paniculata		
*Eycalyptus robusta		
*Eucalyptus saligna		

*Eucalyptus torrelliana		
*Gmelina arborea	Gmeliana	Melina
Grewia mollis		Kyapotro, Sapotoro
Helianthus annus	Sunflower	
Hibiscus spp	Garden hibiscus	
Khaya senegalensis		Dubini (Sarem)
Mangifera indica	Mango	Amango
Parkia clappertoniana		Odonkoran
Parkia bicolor		Odonkoran
Pterocarpus erinaceus		Krayie
Spathodia companulata		Aninsu, Osisiri
Tamarindus indica		Oson
Terminalia ivorensus	Emire	Amire
Terminalia superba		Ofram
Vitex doniana		Afetewa

*Exotic tree

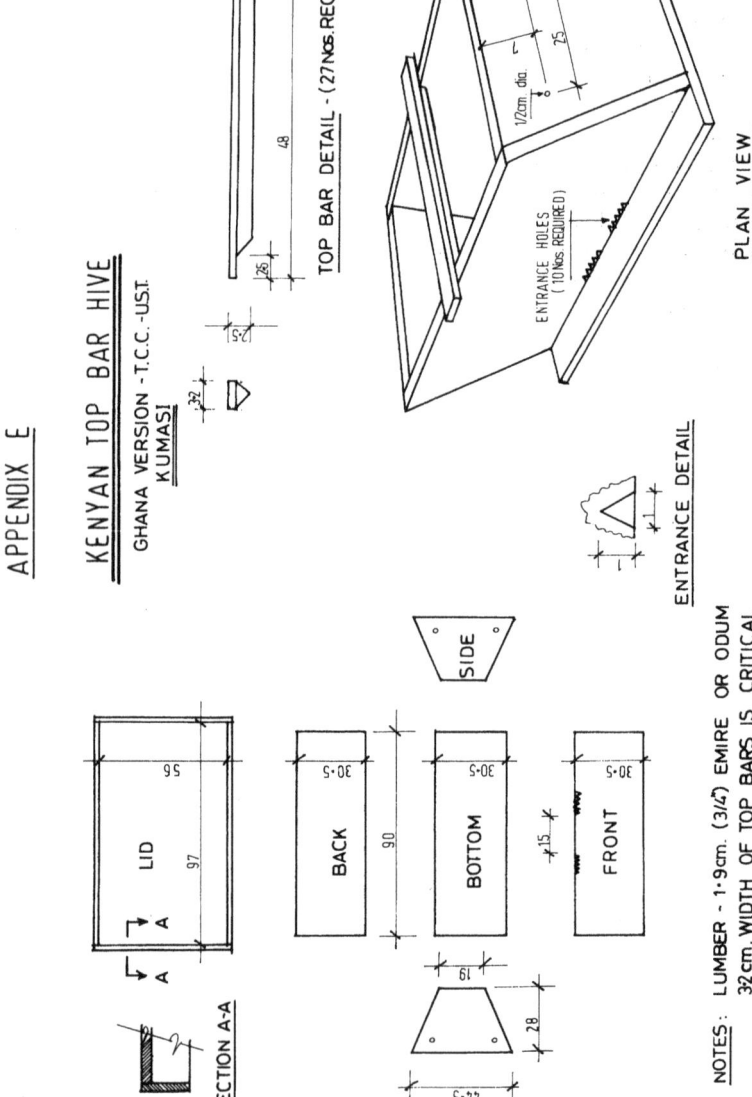

www.ingramcontent.com/pod-product-compliance
Ingram Content Group UK Ltd.
Pitfield, Milton Keynes, MK11 3LW, UK
UKHW041914140426
5217IPUK00013B/150